Adaptive Control
Approach for
Software Quality
Improvement

T0324976

SERIES ON SOFTWARE ENGINEERING AND KNOWLEDGE ENGINEERING*

Editor-in-Chief: **S K CHANG** (*University of Pittsburgh, USA*)

*For the complete list of titles in this series, please go to
http://www.worldscibooks.com/series/ssekes_series

Series on Software Engineering and Knowledge Engineering **Vol. 20**

editors

W Eric Wong

University of Texas at Dallas, USA

Bojan Cukic

West Virginia University, USA

Adaptive Control Approach for Software Quality Improvement

World Scientific

NEW JERSEY · LONDON · SINGAPORE · BEIJING · SHANGHAI · HONG KONG · TAIPEI · CHENNAI

Published by

World Scientific Publishing Co. Pte. Ltd.

5 Toh Tuck Link, Singapore 596224

USA office: 27 Warren Street, Suite 401-402, Hackensack, NJ 07601

UK office: 57 Shelton Street, Covent Garden, London WC2H 9HE

British Library Cataloguing-in-Publication Data
A catalogue record for this book is available from the British Library.

ADAPTIVE CONTROL APPROACH FOR SOFTWARE QUALITY IMPROVEMENT
Series on Software Engineering and Knowledge Engineering — Vol. 20

Copyright © 2011 by World Scientific Publishing Co. Pte. Ltd.

ISBN-13 978-981-4340-91-5
ISBN-10 981-4340-91-X

Typeset by Stallion Press
Email: enquiries@stallionpress.com

Printed in Singapore.

PREFACE

The expansion of our reliance on software in many aspects of modern society has coincided with a number of incidents in aeronautics, astronautics, transportation, medical devices, energy generation, banking and finance. Failures caused by software have introduced more than just inconvenience, but significant property damage, monetary loss, or even fatalities. Therefore, it is of utmost importance that software systems achieve their expected level of quality. As systems grow in complexity, building software free of failure becomes more and more difficult. Some of the most challenging and promising research topics include self management and adaptation at run time, responding to changing user needs and environments, faults, and vulnerabilities. It is critical for researchers and practitioners to understand how these challenges can be addressed to produce high quality software more effectively and efficiently. Control theoretic approaches described in this book represent state-of-the-art techniques that provide some of the answers to these challenges.

Specialized books on the topic of *Software Quality* typically emphasize improvements in various phases of the software development lifecycle, ranging from requirements, architecture, design, implementation, testing, debugging, maintenance, etc. The concept of control theory has been introduced into software engineering recently to analyze online evolution and adaptation of software behavior, to meet old and new functional and non-functional objectives in the presence of changes in the environment, disturbances, faults, or expanded requirements. Due to the novelty of this subject, books on software engineering or control theory have not covered it with a sufficient level of detail.

To overcome such a problem, this book focuses on how adaptive control approach can be applied to improve the quality of software systems. It addresses the following issues:

(1) The application of control theory principles to software processes and systems.

(2) Formalization and quantification of feedback and self-adaptive control mechanisms in software quality assurance.

(3) Integration and the interplay between software engineering and control engineering theory.

Diverse research topics (such as requirements engineering, software development processes, service-oriented architectures, online adaptation of software behavior, testing and QoS control) are woven together into a coherent whole.

Written by world-renowned experts, this book gives an authoritative reference for students, researchers and practitioners to better understand how the adaptive control approach can be applied to improve the quality of software systems. In addition, each chapter outlines future theoretical and experimental challenges for researchers in this area.

We would like to thank all the chapter authors for sharing their ideas and research results with us, and the reviewers for their valuable feedback on the chapters they reviewed. This book would not be possible without their contributions. Special thanks go to Professor Shi-Kuo Chang, Editor-in-Chief of the Book Series on *Software Engineering and Knowledge Engineering*, for approving this project as well as his continuous support and encouragement; Mr. Andrew Restrepo, a PhD student in Computer Science at the University of Texas at Dallas, for editing earlier drafts of this book; and Mr. Steven Patt, our production editor, and other staff at World Scientific Publishing in assisting with all the logistics during the development of this book.

We are grateful to our families, Nancy Chen, Rachel Wong, Justin Wong, and Vincent Wong, and April Cukic. Without their love and support this project could not have been accomplished.

W. Eric Wong & Bojan Cukic
April 15, 2011

CONTENTS

Chapter 1

PRIORITIZING COVERAGE-ORIENTED TESTING PROCESS — AN ADAPTIVE-LEARNING-BASED APPROACH AND CASE STUDY

FEVZI BELLI

Department of Computer Science, Electrical Engineering and Mathematics
Institute for Electrical Engineering and Information Technology
University of Paderborn
Warburger Straße 100 D-33098 Paderborn, Germany
belli@upb.de

MUBARIZ EMINOV

Halic University, Faculty of Engineering
Department of Computer Engineering, Istanbul. Turkey
mubarizeminli@halic.edu.tr

NIDA GÖKÇE

Department of Statistics, Faculty of Science
Mugla University, 4800 Mugla, Turkey
gnida@mu.edu.tr

W. ERIC WONG

Department of Computer Science, University of Texas at Dallas
Richardson, Texas 75080, USA
ewong@utdallas.edu

This chapter proposes a graph-model-based approach to prioritizing the test process. Tests are ranked according to their preference degrees which are determined indirectly, i.e., through classifying the events. For construction of the groups of events, an unsupervised neural network is trained by adaptive competitive learning algorithm. A case study demonstrates and validates the approach.

1. Introduction and Related Work

Testing is one of the important, traditional analytical techniques of quality assurance in the software industry. There is no justification, however, for any assessment of the correctness of system under test (SUT) based on the success of a single test, because potentially there can be an infinite number of test cases. To overcome this principal shortcoming of testing concerning completeness of the validation, formal methods have been proposed. Those models visualize and represent the relevant, desirable features of the SUT.

For a productive generation of tests, model-based techniques focus on particular, relevant aspects of the requirements of the SUT and its environment. Real-life SUTs have, however, numerous features that are to simultaneously be considered, often leading to a large number of tests. In many applications where testing is required, the complete set of tests is not run due to time or budget constraints. In such cases, the entire set of system features cannot be considered. In these situations, it is essential to *prioritize* the test process. It is then essential to model the relevant features of SUT. The modeled features are either functional behavior or structural issues of the SUT (as given in its code), leading to *specification-oriented* testing or *implementation-oriented* testing, respectively. Once the model is established, it "guides" the test process to generate and select test cases, which form sets of test cases (also called *test suites*). The test selection is ruled by an *adequacy criterion*, which provides a measure of how effective a given set of test cases is in terms of its potential to reveal faults.[1,2] Some of the existing adequacy criteria are coverage-oriented. They use the ratio of the portion of the specification or code that is covered by the given test set in relation to the uncovered portion in order to determine the point in time at which to stop testing (*test termination problem*).

Test case prioritization techniques organize the test cases in a test suite by ordering such that the most beneficial are executed first thus allowing for an increase in the effectiveness of testing. One of the performance goals, i.e., the fault detection rate, is a measure of how quickly faults are detected during the testing process.[3]

This chapter is on model-based, specification- and coverage-oriented testing. The underlying model graphically represents the system behavior interacting with the user's actions. In this context, *event sequence graphs* (ESG)[9-11] are favored. ESG approach view the system's behavior and user's actions as events, more precisely, as *desirable events*, if they are in accordance with the user expectations, otherwise they are *undesirable events*. Mathematically speaking, a complementary view of the behavioral model is generated from the model given. Thus, the model will be exploited twice, i.e., once to validate the system behavior under regular conditions and a second time to test its robustness under irregular, unexpected conditions.

The costs of testing often tend to run out the limits of the test budget. In those cases, the tester may request a complete test suite and attempt to run as many tests as affordable, without running out the budget. Therefore, it is important to test the most important items first. This leads to the *Test Case Prioritization Problem (TCPP)* a formal definition of which is represented in[5] as follows:

Given: A test suite T; the set PT of permutations of T; a function f from PT to the real numbers which represents the preference of the tester while testing.

Problem: Find $T' \in PT$ such that $(\forall T'')\,(T'' \neq T')\,[f(T') \geq f(T'')]$

Existing approaches to solving TCPP usually suggest constructing a density covering array in which all pair-wise interactions are covered.[4,5] Generally speaking, every n-tuple is then qualified by a number $n \in \mathbb{N}$ ($\mathbb{N}$: set of natural numbers) of values to each of which a degree of importance is assigned. In order to capture significant interactions among pairs of choices the importance of pairs is defined as the "benefit" of the tests. Every pair covered by the test contributes to the total benefit of a test suite by its individual benefit. Therefore, the tests given by a test suite are to be ordered according to the importance of corresponding pairs. However, such interaction-based, prioritized algorithms are computationally complex and thus usually less effective.[6,7]

The ESG approach favored in this chapter generates test suites through a finite sequence of discrete events. The underlying optimization problem is a generalization of the *Chinese Postman Problem (CPP)*[8] and algorithms given in[9-11] differ from the well-known ones in that they satisfy not only the constraint that a minimum total length of test sequences is required, but also fulfill the coverage criterion with respect to converging of all event pairs represented graphically. This is substantial to solve the test termination problem and accounts for a significant difference of this present chapter from existing approaches. To overcome the problem that an exhaustive testing might be infeasible, the present chapter develops a *prioritized* version of the mentioned test generation and optimization algorithms, in the sense of "divide and conquer" principle. This is the primary objective and the kernel of this chapter which is novel and thus, to our knowledge, has not yet been worked out in previous work.

The required prioritization has to meet the needs and preferences of test management on how to spend the test budget. However, SUT and software objects, i.e., components, architecture, etc., usually have a great variety of features. Therefore, test prioritization entails the determination of order relation(s) for these features. Generally speaking, we have n objects, whereby each object has a number (p) of features that we call *dimension*.

TCPP then represents the comparison of test objects of different, multiple dimensions. To our knowledge, none of the existing approaches take the fact into account that SUT usually has a set of attributes and not a single one when prioritizing the test process. Being of enormous practical relevance, this is a tough, *NP*-complete problem.

Our approach assigns to each of the tests generated a degree of its preference. This degree is indirectly determined through estimation of the events qualified by several attributes. We suggest representing those events as an unstructured multidimensional data set and dividing them into groups which correspond to their importance. Beforehand, the optimal number of those groups is determined by using V_{sv} *index-based clustering validity algorithm.*[12,13] To derive the groups of events we use a clustering approach based on unsupervised neural networks (NN) that will be trained by an adaptive competitive learning (CL) algorithm.[14] Different from the existing approaches, e.g., as described in,[12,15,16] input and weight vectors are normalized, i.e., they have length one. This enables less sensitivity to initialization and a good classification performance. The effectiveness of the proposed testing approach is demonstrated and validated by a case study a non-trivial commercial system.

The chapter is organized as follows. Section 2 explains the background of the approach, presenting also the definition of neural network-based clustering. Section 3 explains the CL algorithms. Section 4 describes the proposed prioritized graph-based testing approach. Section 5 includes the case study. Section 6 summarizes the results, gives hints to further research and concludes the chapter.

2. Background

2.1. *Event Sequence Graphs*

Because the construction of ESG, test generation from ESG and test process optimization are sufficiently explained in the literature,[9–11,17] the present chapter summarizes ESG concept, as far as it is necessary and sufficient to understand the test prioritization approach represented in this chapter.

Basically, an *event* is an externally observable phenomenon, such as an environmental or a user stimulus, or a system response, punctuating different stages of the system activity. A simple example of an ESG is given in Fig. 1. Mathematically, an ESG is a directed, labeled graph and may be thought of as an ordered pair $ESG = (\alpha, E)$, where α is a finite set of

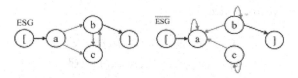

Fig. 1. An event sequence graph ESG, its complement $\overline{ESG}$.

nodes (vertices) uniquely labeled by some input symbols of the alphabet Σ, denoting events, and E: $\alpha \rightarrow \alpha$, a precedence relation, possibly empty, on α. The elements of E represent directed arcs (edges) between the nodes in α. Given two nodes a and b in α, a directed arc ab from a to b signifies that event b can follow event a, defining an *event pair* (*EP*) ab (Fig. 1). The remaining pairs given by the alphabet Σ, but not in the ESG, form the set of *faulty event pairs* (*FEP*), e.g., ba. As a convention, a dedicated, start vertex, e.g., [, is the *entry* of the ESG whereas a final vertex e.g.,] represents the *exit*. Note that [and] are not included in Σ; therefore, the arcs from and to them form neither EP nor FEP. The set of FEPs constitutes the *complement* of the given ESG ($\overline{ESG}$). Superposition of ESG and $\overline{ESG}$ leads to completed ESG ($\widehat{ESG}$) (Fig. 1).

A sequence of $n + 1$ consecutive events that represents the sequence of n arcs is called a *event sequence* (*ES*) *of the length* $n + 1$, e.g., an *EP* (*event pair*) is an ES of length 2. An ES is *complete* if it starts at the initial state of the ESG and ends at the final event; in this case it is called a *complete ES* (*CES*). Occasionally, we call CES also *walks* (or *paths*) through the ESG given. A *faulty event sequence* (*FES*) *of the length* n consists of $n - 1$ subsequent events that form an ES of length $n - 2$ plus a concluding, subsequent FEP. An FES is *complete* if it starts at the initial state of the ESG; in this case it is called *faulty complete ES*, abbreviated as *FCES*. A FCES must not necessarily end at the final event.

2.2. *Neural Network-Based Clustering*

Clustering is a technique to generate an optimal partition of a given, supposedly unstructured, data set into a predefined number of *clusters* (or *groups*). Homogeneity within the groups and heterogeneity between them can be settled by means of unsupervised neural network-based *clustering algorithms*.[13,14] For clustering of an unstructured data set dealing especially with vector quantization, unsupervised learning based on clustering in a neural network framework is frequently used. Clustering has to obtain partition data vector space

$$X = \{x_1, \ldots, x_i, \ldots, x_n\} \subset \mathbb{R}^p,$$
$$x_i = (x_{i1}, \ldots, x_{ij}, \ldots x_{ip}) \in \mathbb{R}^p \tag{1}$$

into c number clusters or subspaces S_k in the form of hyper spherical clouds of pattern vectors $x_i = \{x_1, x_2, \ldots, x_p\} \in R^p$. Each of these subspaces is represented by a cluster center (prototype) that corresponds to weight vector $w = (w_1, w_2, \ldots, w_p) \in R^p$. An input (pattern) vector x_i is described by the best-matching or "winning" weight vector w_k for which criterion-distortion error $d(x_i, w_k) = \|x_i - w_k\|^2$ that is the squared error of Euclidean distance is minimum. The procedure divides the input space R^p into partition subspace

$$S_k = \{X \in R^p \,\|\|x - w_k\| \le \|x - w_j\| \,\forall j \neq k\} \quad k = 1, \ldots, c \in \mathbb{N} \tag{2}$$

called Voronoi polygons or Voronoi polyhedra. To determine the optimal number c of groups, the V_{sv} *index-based cluster validity algorithm*[13] has been used.

To provide *training* of NN, a number of *learning algorithms* are used. We deal with the family of *competitive learning* (CL) algorithms[14] that are a type of self-organizing networking models. According to CL algorithms, "winning" weight vector (weights of connections between input and output nodes) or *cluster center* can be adjusted by applying "winner-takes-all" strategy in training phase of the NN under consideration. In clustering a data set which is to be partitioned into c number of clusters each of which contains a data subset S_k defined as follows:

$$X = \bigcup_{k=1}^{c} S_k \quad \text{with } S_k \cap S_j = 0 \,\,\forall k \neq j \tag{3}$$

An optimal partition $x_i \in R^p$ into subspaces S_k, $k = 1, 2, \ldots, c$, is obtained through an optimal choice of reference vectors w_k which minimize a cost function-distortion error represented as follows:

$$E = \sum_{k=1}^{c} \int_{S_k} d(x, w_k) g(x) dx \tag{4}$$

where $g(x)$ is a probability density function. If probability distribution of data vectors $g(x)$ is known in advance then gradient descent algorithm can be applied on E in order to minimize (4) which leads to well-known k-means clustering algorithm[18] under fixed number of subspaces (clusters). Thus, optimal weight vectors w_k, $k = 1, \ldots, c$, can be precisely determined. However, in general, $g(x)$ is not given a priori, therefore a number of neural network clustering algorithms[15,16,19,20] were suggested to evaluate this unknown density function. In the case, when the number of the clusters c increases, density function $g(x)$ in each cluster becomes approximately uniform,[27] therefore (4) can be rewritten as

$$E = \sum_{k=1}^{c} g(w_k) \int_{S_k} d(x, w_k) dx \qquad (5)$$

Let D_k be partition error in k-th subspaces S_k, then

$$E = \sum_{k=1}^{c} D_k \qquad (6)$$

where $D_k = g(w_k) \int_{S_k} d(x, w_k) dx$. As the sequence of input vectors becomes stationary and ergodic, it is known that (5) is corresponding to (7) presented in the mean square error (MSE) as follows[16,19]

$$E = \frac{1}{n} \sum_{k=1}^{c} D_k \qquad (7)$$

where

$$D_k = \frac{1}{p} \left(\sum_{x \in S_k} d(x, w_k) \right) \qquad (8)$$

n is the total number of input vectors and p is the dimension of input vector. Therefore, the optimal clustering results in obtaining partition subspaces S_k and weight vectors w_k, $k = 1, 2, \ldots, c$ that minimize D_k.

3. Competitive Learning

CL is a paradigm in which a structured or unstructured population of units compete with each other with regard to a stimulus, where the winner (or winners) of the competition may respond and be adapted. Algorithm that implements CL is suited to different specific concerns, although it is generally nonparametric, and suited to the general domains of function approximation, classification, and regression.[21]

CL is a connectionist machine learning paradigm where an input pattern is matched to the node with the most similar input weights, and the weights are adjusted to better resemble the input pattern. This is called the *winner-take-all* (or maximum activation) unsupervised learning method where the input pattern is compared to all nodes based on similarity. The nodes compete for selection (or stimulation) and ultimately adjustment (or learning).[22] Kohonen distinguishes this connectionist learning paradigm from feed-forward and feed-backward approaches,[23,24] as follows:

Signal Transfer Networks: (feed-forward paradigm) Signal transform circuits where the output signals depend on the input signals received by the network. Parametric in that the mapping is defined by a basis function (components of the structure) and fitted using an optimization approach like gradient decent. Examples include the multilayer Perceptron, back propagation, and radial basis function.

State-Transfer Networks: (feed-backward paradigm) Based on relaxation effects where the feedbacks and nonlinearities cause the activity state to quickly converge to one of its stable values (attractor). Input signals provide the initial activity state, and the final state is a result of recurrent feedbacks and computation. Examples include Hopfield network, Boltzmann machine, and bidirectional associative memory (BAM).

Competitive Learning: (self-organizing network paradigm) Networks of cells in simple structures receive identical inputs from which they compete for activation through positive and negative lateral interactions. One cell is the winner, and other cells are inhibited or suppressed. Cells become sensitive to different inputs and act as decoders for the domain. The result is a globally ordered map created via a self-organizing process. Examples include the Self-Organizing Map (SOM), and Learning Vector Quantization (LVQ).[21]

Fritzke[25] uses taxonomy of hard (*winner-take-all*) and soft (*winner-take-most*) CL and further distinguishes soft approaches to those with and without a fixed network topology.

Hard CL: Winner-take-all (WTA) learning each input signal results in the adaptation of a single unit of the model. These methods may occur online or offline in batch. Examples include *k*-means.

Soft CL: Winner-take-most (WTM) learning where an input signal results in the adaptation of more than one unit of the model. No fixed model dimensionality or topology is prescribed with these methods. Examples include neural gas.

Soft CL with Fixed Structure: Winner-take-most (WTM) learning with a fixed model dimensionality and or topology. Examples include the self-organizing map.

3.1. *Distance-Based Competitive Learning Algorithm*

CL is closely related to clustering that learns to group input patterns in clusters. In the input space $x \in R^p$ the input pattern x_i is defined, which is generated upon the probability density function $g(x)$, by applying the winner-take-all strategy. When both input vectors and weight vectors are not normalized, the Euclidean distance measure, in general, is used to determine the winner weight vector w_w in CL algorithms.

Training: In the *training* phase the weight vectors of NN are updated usually according to Standard CL algorithm. Firstly, for a data point $x_i \in \mathbb{R}^p$ selected from X the winner weight vector w_w is determined by:

$$w_w = \arg\min_k \{\|x_i - w_k\|\}$$
$$i = 1, \ldots, n \in \mathbb{N} \quad k = 1, \ldots, c \in \mathbb{N} \tag{9}$$

where $\|.\|$ is *Euclidean distance measure*. Then this vector is adjusted at step t by

$$\Delta w_w(t) = \eta(t)(x_i - w_w) \tag{10}$$

where $\eta(t)$ is a *learning rate*. Secondly, the adjusted winner vector is calculated by

$$w_w(t) = w_w(t-1) + \Delta w_w(t) \tag{11}$$

Training process iteratively proceeds until the convergence condition for the all weight vectors is satisfied. Clearly, the CL algorithm actually seeks for a local minimum (with respect to the predetermined number of clusters) for squared error criterion by applying gradient descent optimization. As known, in the Kohonen's SOFM algorithm not only w_w but also the weight vectors that are placed in its neighborhood are adjusted. The learning rate for these weight vectors is set to be much smaller than the rate for w_w that is slowly reduced up to the winner weight vector. Thus, the updating rule of this learning algorithm becomes as

$$\Delta w_k(t) = \eta(t)(x_i - w_k) \quad k \in N_c \tag{12}$$

where $N_c(t)$ has a set of indexes of neighborhoods for the winner w_k at step t. If $N_c(t)$ has index of the winner only, then Kohonen's algorithm becomes the standard CL algorithm presented in above.

3.2. Angle-Based Competitive Learning Algorithm

Now, we consider the updating rule of CL algorithm when both input vectors and the weight vectors are normalized to a unit length, that is, all vectors are presented as the unit vectors. For the input patterns $x \in R^p$, the corresponding normalized vectors $\tilde{x}$ are given by

$$\tilde{x} = \frac{x}{|x|} = x \cdot \left(\sum_{j=1}^{p} x_j^2 \right)^{-1/2} \tag{13}$$

where $|x|$ is the magnitude of input vector x that lies on a unit hyper sphere in R^p. In this case, as known, the winner weight vector is determined by the dot product of the presented input vector x_i and a weight vector w_k, then (9) can be reformed as the follows

$$\tilde{w}_w = \arg\max_k \left\{ \sum_{j=1}^{p} \tilde{x}_{ij} \tilde{w}_{kj} \right\}$$
$$i = 1, \ldots, n \quad k = 1, \ldots, c \tag{14}$$

i.e., the winner vector $\tilde{w}_w$ is chosen by the largest activation level. Since the dot product is $\cos\theta$ where θ the angle between is two considered vectors, then (14) can be expressed as

$$\tilde{w}_w^\theta = \arg\min_k \{\theta_k\} \quad k = 1, \ldots, c \tag{15}$$

i.e., the winner vector $\tilde{w}_w^\theta$ is determined by the smallest angle level between the presented x_i and weight vectors w_k, $k = 1, \ldots, c$. The updating rule of a winner weight vector instead of (10) is based on the adjusting equation (16) expressed as follows

$$\Delta \tilde{w}_w(t) = \eta(t) \left(\frac{\tilde{x}_i}{p} - \tilde{w}_w \right) \tag{16}$$

Then for Kohonen's SOFM algorithm the updating rule has the following form:

$$\Delta \tilde{w}_k(t) = \eta(t) (\tilde{x}_i - \tilde{w}_k) \quad k \in N_c \tag{17}$$

Thus, the winner weight vector at step t will be

$$\tilde{w}_w(t) = \tilde{w}(t-1) + \Delta\tilde{w}_w(t) \tag{18}$$

However, in general, the distribution is not given in advance; hence the initial values of the weight vectors are randomly allotted. It negatively influences the clustering performance of the considered CL algorithm.

3.3. *Adaptive Competitive Learning*

In this section we present the CL algorithm for neural network clustering, which is able to: have limited dependence on initial values of weight vectors; reduce partition performance. Similar to the studies in[15,16] disclosed above, it uses the deletion method that eliminates sequentially weight vectors which are prepared more than their predetermined numbers in advance. As in[9] where this learning algorithm is called the adaptively CL, we use the simplest standard CL algorithm. However, instead of direct employing of the input vectors, it utilizes the corresponding vectors normalized to a unit length. Rummelhart[26] introduced such kind of normalization of the vectors in input space for CL and afterwards it was used in the few versions of Kohonen's SOFM algorithm.[20] It has been utilized for classification, as well as shown good performance.[27]

In the suggested CL algorithm we use a deletion method based on a criterion of subdistortion or intra-cluster partition error but in this case its equation will be different from (8) and it becomes as

$$D_k = \frac{1}{p}\left(\sum_{\tilde{x} \in S_k} \tilde{x}\tilde{w}_w\right) \quad k = 1, \ldots, c \tag{19}$$

The self-elimination procedure carried out according to (19) is shortly described as the follows. After learning by standard CL algorithm, a weight vector w_s that has a minimum intra-cluster partition error, i.e., $D_s \geq D_k$, for all k, is deleted. Due to the use of the activation-based selection of the winner vector we call the suggested algorithm as the activation checking based CL with deletion. In this case, to signify subdistortion error as minimal we use the angle estimation of D_k presented as the follows

$$D_k^* = \frac{1}{p}\left(\sum_{\tilde{x} \in S_k} \theta_k\right) \quad k = 1, \ldots, c \tag{20}$$

Then using (20) and (7), clustering performance can be estimated by criteria such as the mean angle error and the standard angle deviation among subdistortions for given data set. Thus, the proposed Adaptive CL algorithm is presented as follows.

Adaptive Competitive Learning Algorithm:

Step 1. Initialization:

Initial number of output neurons l_0, final number of neurons l, maximum iteration T_{max}, initial iteration of deletion $t_0 = T_{max}/3$ and partial iteration $u = T_{max}/3(l_0 - l + 1)$, Set $t \leftarrow 0$ and $m \leftarrow l_0$

Step 2. Angle-Based Competitive Learning:

2.1. Choose an input vector $\tilde{x}_i$ at random among X

2.2. Select a winner $\tilde{w}_k$ according to (14)

2.3. Update the winner $\tilde{w}_w$ vector according to (16)

2.4. Set $t \leftarrow t + 1$

2.5. If $m > l$ and $t = t_0 + u \times q$ than go to Step 3, otherwise go to 2.1.

Step 3. Deletion Mechanism:

3.1. Delete $\tilde{w}_k$ calculating D_k according to (19) and checking $D_s \geq D_k$

3.2. Set $m \leftarrow m - 1$

Step 4. Termination Condition:

If $t = T_{max}$ then terminate, otherwise go to Step 2.

Classification: After finding a value of weight vectors $\{w_1, \ldots, w_c\}$ that correspond to cluster centers, respectively, a data set is divided into c groups as follows:

$$S_k = \left\{ x \in \mathbb{R}^p \,\middle|\, \sum_{j=1}^{p} \tilde{x}_{ij}\tilde{w}_{kj} \geq \sum_{j=1}^{p} \tilde{x}_{ij}\tilde{w}_{mj} \forall k \neq m \right\} \qquad (21)$$

$$i = 1, \ldots, n, \quad j = 1, \ldots, p, \quad k = 1, \ldots, c, \quad m = 1, \ldots, c \in \mathbb{N}$$

Classification performance of the considered clustering algorithm was estimated by the MSE calculated using (7) and (19). Effectiveness of this algorithm was verified for different types of data sets in.[14] Computational time for classification depends on the number n of the events and the number p of the attributes.

4. Prioritized ESG-Based Testing

We consider the testing process based on the generation of a test suite from ESG that is a discrete model of a SUT. To generate tests, firstly a set of ESGs are derived which are input to the generation algorithm to be applied. We deal with the test generation algorithms[9-11] that generates tests for a given ESG and satisfies the following coverage criteria.

(a) Cover all event pairs in the ESG.
(b) Cover all faulty event pairs derived by the $\overline{ESG}$.

Note that a test suite that satisfies the first criterion consists of CESs while a test suite that satisfies the second consists of FCESs. These algorithms are able to provide the following constraints:

(a) The sum of the lengths of the generated CESs should be minimal.
(b) The sum of the lengths of the generated FCESs should be minimal.

The constraints on total lengths of the tests generated enable a considerable reduction in the cost of the test execution and thus the algorithms mentioned above can be referred to as the relatively efficient ones. However, as stated in Section 1, an entire test suite generated may not be executed due to limited project budget. Such circumstances entail ordering all tests to be checked and exercised as far as they do not exceed the test budget. To solve the test prioritizing problem, several algorithms have been introduced.[1,4] Usually, during the test process for each *n-tuple* (in particular pair-wise) interaction a degree of importance is computationally determined and assigned to the corresponding test case. However, this kind of prioritized testing is computationally complex and hence restricted to deal with short test cases only.

Our prioritized testing approach is based on the ESG-based testing algorithms mentioned above. Note that our test suite consists of CESs which start at the entry of the ESG and end of its exit, representing *walks* (*paths*) through the ESG under consideration. This assumption enables to order the generated tests, i.e., CESs.

The ordering of the CESs is in accordance with their preference degree which is defined indirectly, i.e., by estimation of events that are the nodes of ESG and represent objects (modules, components) of SUT. For this aim, firstly events are presented as a *multidimensional event vector* $x_i = (x_1, \ldots, x_p)$ where p is the *number of attributes*.

4.1. *Definition of the Attributes of Events*

To qualify an event corresponding to a node in ESG, as a arbitrarily chosen example, we propose to use following 9 attributes, i.e., $p = 9$, that determine the *dimension* of a data point represented in a data set.[28,29] These attributes are given below:

x_1: The number of sub-windows to reach an event from the entry [(gives its distance to the beginning).

x_2: The number of incoming and outgoing edges (invokes usage density of a node, i.e., an event).

x_3: The number of nodes (events) which are directly and indirectly reachable from an event except entry and exit (indicates its "traffic" significance).

x_4: The maximum number of nodes to the entry [(its maximum distance in terms of events to the entry).

x_5: The number of nodes (events) of a sub-node as sub-menus that can be reached from this node (maximum number of sub-functions that can be invoked further).

x_6: The total number of occurrences of an event (a node) within all CESs, i.e., walks (significance of an event).

x_7: The *balancing degree* determines balancing a node as the sum of all incoming edges (as plus $(+)$) and outgoing edges (as minus $(-)$) for a given node.

x_8: The averaged frequencies of the usage of events within the CESs (determines the averaged occurrence of each event within all CESs).

x_9: The number of FEPs connected to the node under consideration (takes the number of all potential faulty events entailed by the event given into account).

4.2. *Definition of Importance Degree and Preference*

The CESs are manually ordered, scaling their preference degrees based on the events which incorporate the importance group(s). *Importance* (Imp(e)) of eth event is defined as follows[28]:

$$Imp(e) = c - ImpD(S_k) + 1 \qquad (22)$$

where c is the optimal number of the groups; $ImpD(S_k)$ is defined by means of the importance degree of the group S_k to which the eth event belongs.

Finally, choosing the events from the ordered groups, a ranking of CESs is formed according to their descending preference degrees. The assignment of preference degrees to CESs is based on the following rule:

(a) The CES under consideration has the highest degree if it contains the events which belong to the "top" group(s) with utmost importance degrees, i.e., that is placed within the highest part of group ordering.

(b) The CES under consideration has the lowest degree if it contains the events which belong to the group(s) that are within the lowest part of the "bottom" group(s) with least importance degree i.e., that is placed within the lowest part of group ordering.

Therefore, the preference degree of CES can be defined by taking into account both the importance of events (22) and the frequency of occurrence of event(s) within them that is formulated as follows[24]:

$$\mathrm{Pref}(\mathrm{CES}_q) = \sum_{e=1}^{n} \mathrm{Imp}(e) f_q(e) \quad q = 1, \ldots, m \in \mathbb{N} \qquad (23)$$

where m is the number of CESs, n is the number of events, $\mathrm{Imp}(e)$ is importance degree of the eth event (22) and $f_q(e)$ is frequency of occurrence of event e within CES_q. This order determines the *preference degree* ($\mathrm{Pref}(\mathrm{CES}_q)$) of CESs as test cases (23).

Indirect Determination of the Preference Degree

Step 1. Construction of a set of events $X = \{x_{ij}\}$ where $i = 1, \ldots, n \in \mathbb{N}$ is an event index, and $j = 1, \ldots, p \in \mathbb{N}$ is an attribute index.

Step 2. Training the NN using adaptive CL algorithm (see Section 3.2).

Step 3. Classification of the events into c groups ((21), see Section 3.2).

Step 4. Determination of importance degrees of groups according to length (ℓ) of weight vectors.

Step 5. Determination of importance degrees of event groups ((22), this present section).

Step 6. An ordering of the CESs for prioritizing the test process.

5. A Case Study

Based on the web-based system ISELTA (*Isik's System for Enterprise-Level Web-Centric Tourist Applications*), we now present a case study to validate the testing approach presented in the previous sections.[15] Both the construction of ESGs, and generation of test cases from those ESGs, have

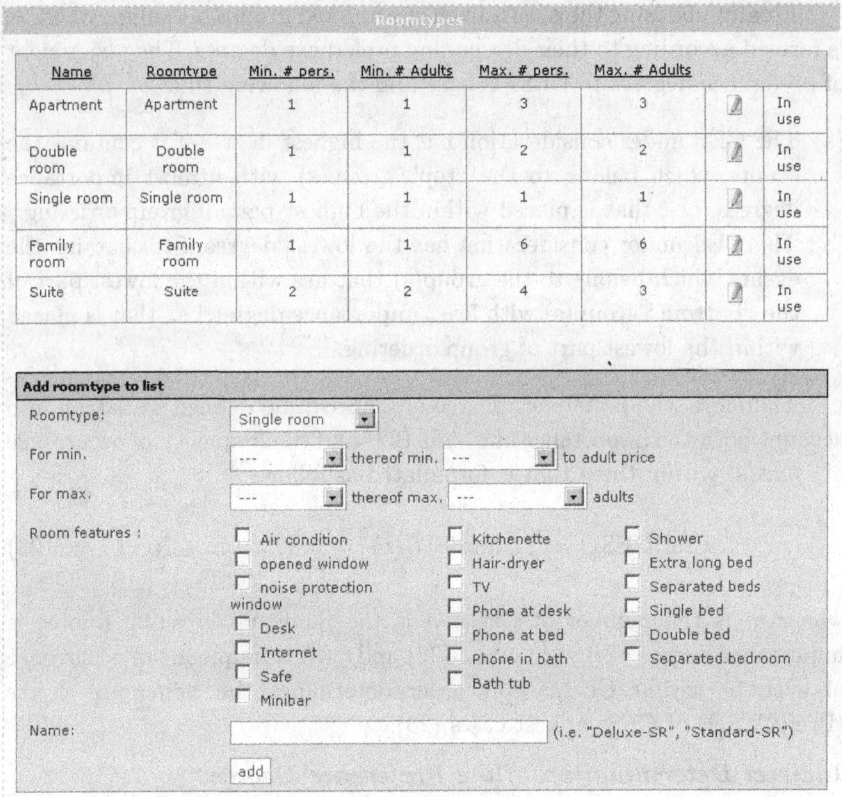

Roomtypes

Name	Roomtype	Min. # pers.	Min. # Adults	Max. # pers.	Max. # Adults		
Apartment	Apartment	1	1	3	3		In use
Double room	Double room	1	1	2	2		In use
Single room	Single room	1	1	1	1		In use
Family room	Family room	1	1	6	6		In use
Suite	Suite	2	2	4	3		In use

Add roomtype to list

Roomtype: Single room

For min. --- thereof min. --- to adult price

For max. --- thereof max. --- adults

Room features : Air condition, opened window, noise protection window, Desk, Internet, Safe, Minibar, Kitchenette, Hair-dryer, TV, Phone at desk, Phone at bed, Phone in bath, Bath tub, Shower, Extra long bed, Separated beds, Single bed, Double bed, Separated bedroom

Name: (i.e. "Deluxe-SR", "Standard-SR")

add

Fig. 2. Room definition/reservation process in ISELTA.

been explained in the previous papers of the first author.[9–11] Therefore, the case study concentrates on the test prioritizing problem.

ISELTA has been developed by our group in cooperation with a commercial enterprise to market various tourist services for traveling, recreation and vacation. It can be used by hotel owners, travel agents, etc., but also by end consumers. A screenshot in Fig. 2 demonstrates how to define and reserve rooms of different types.

5.1. *Derivation of the Test Cases*

Figure 3 depicts the completed ESG of the scenario described above and in Fig. 2. Test cases can now be generated using the algorithms mentioned in Section 3 and described in[10–11] in detail. For the lack of space, reference is made to these papers and the CESs generated are listed in Table 1.

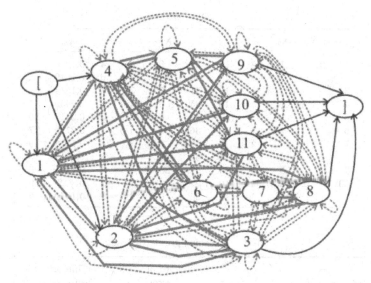

Fig. 3. Completed ESG for room definition/selection (solid arcs: event pairs (EP); dashed.

Legend of Fig. 3:

1: Click on "Starting"
2: Click on "Registering"
3: Registration carried out
4: Click on "log in"
5: Logged in
6: Click on "Password forgotten"

7: Password forgotten
8: Click on "Request"
9: Indicate service(s) offered
10: Indicate administrator
11: Indicate agent

Table 1. CESs of ESG in Fig 3.

CES$_1$: [4 5 4 5 9 1 4 5 10 1 4 5 11 1 4 5 9 2 3 4 5
10 2 3 2 3 1 4 5 11 2 3 4 6 4 6 7 8 1 2 3]
CES$_2$: [1 4 5 9]
CES$_3$: [2 3 4 6 7 8 2 3 4 5 10]
CES$_4$: [4 5 11]
CES$_5$: [4 6 7 8]

5.2. *Determination of Attributes of Events*

As a follow-on step, each event, i.e., the corresponding node in the ESG, is represented as a multidimensional data point using the values of all nine attributes as defined in the previous section. Estimating by means of the

Table 2.   Data set of events.

Event No	X_1	X_2	X_3	X_4	X_5	X_6	X_7	X_8	X_9
1	1	8	10	39	0	6	4	0,1860	8
2	1	8	10	40	0	7	6	0,1519	9
3	2	5	10	41	6	7	−3	0,1519	11
4	1	7	10	35	0	14	3	0,2469	7
5	2	5	10	29	0	10	−3	0,2112	9
6	2	3	10	36	0	4	−1	0,2199	7
7	3	2	10	37	0	3	0	0,1218	9
8	4	4	10	38	0	3	−2	0,1218	10
9	3	4	10	17	30	3	−2	0,1494	9
10	3	4	10	22	0	3	−2	0,0698	9
11	3	4	10	30	0	3	−2	0,1911	9

Table 3.   Obtained groups of events.

Groups (3) and (13)	Events	Length of weight vectors (ℓ)	Importance Degree ImpD(S_k)
S_1	9	2,07	2
S_2	3,6,7,8,11	2,04	3
S_3	1,2,4	1,97	4
S_4	5	2,25	1
S_5	10	1,73	5

ESG and $\overline{ESG}$, the values of attributes for all events are determined and the data set $X = \{x_1, \ldots, x_{11}\} \subset \mathbb{R}^9$ is constructed as in Table 2.

5.3. *Construction of the Groups of Events*

For the data set gained from the case study (Fig. 2 and 3), the optimal number c of the groups is determined to be 5 which leads to the groups $S_k, k = 1, \ldots, 5$. Importance degrees ($ImpD(S_k)$) of obtained groups are determined by comparing the length of their weight vectors (ℓ) and all $ImpD(S_k)$ values that are presented in Table 3.

5.4. *Indirect Determination of Preference Degrees*

As mentioned in the previous section, the preference degree of the CESs is determined indirectly by (23) that depend on the importance of events (22) and frequency of event(s) within CES. The ranking of the CESs is represented in Table 4.

Table 4.　Ranking of CESs (walks).

PrefDeg.	Pref (CES$_q$)(15)	CESs	CESs (walks)
1	126	CES$_1$	[4 5 4 5 9 1 4 5 10 1 4 5 11 1 4 5 9 2 3 4 5 10 2 3 2 3 1 4 5 11 2 3 4 6 4 6 7 8 1 2 3]
2	29	CES$_3$	[2 3 4 6 7 8 2 3 4 5 10]
3	13	CES$_2$	[1 4 5 9]
4	11	CES$_5$	[4 6 7 8]
5	10	CES$_4$	[4 5 11]

Exercising the test cases (CESs, or walks) in this order ensure that the most important tests will be carried out first. Moreover, the achieved ranking of CESs complies with the tester's view. Thus, an ordering of the complete set of CESs (walks) is determined using the test suite generated by the test process, i.e., we now have a ranking of test cases to make the decision of which test cases are to be primarily tested. Undesirable events can be handled in a similar way; therefore, we skip the construction of ranking of the FCES.

6. Conclusions and Future Work

The model-based, coverage-and specification-oriented approach described in the previous sections provides a novel and effective algorithm for ordering the test cases according to their degree of preference. Such degrees are determined indirectly through the use of the events specified by several attributes, and not a single one. This is an important issue and consequently, the approach introduced radically differs from the existing ones.

The relevant attributes are visualized by means of a graphical representation (here, given as a set of ESGs). The events (nodes of ESG) are classified using unsupervised neural network clustering. The approach is useful when an ordering of the tests due to restricted budget and time is required. Run-time complexity of this approach is of $o(n^2)$, assuming that the number of events (n) greater than the number of attributes (p), otherwise it is $o(p^2)$.

We plan to apply our prioritization approach to a more general class of testing problems, e.g., to multiple-metrics-based testing where a family of software measures is used to generate tests.[30] Generally speaking, the approach can be applied to prioritize the testing process if the SUT is modeled by a graph of the nodes which represent events or sub-systems of various granularities (modules and functions, or objects, methods, classes, etc.).

References

1. Binder, R.V.: *Testing Object-Oriented Systems.* Addison-Wesley, 2000.
2. Gerhart, S., Goodenough, J.B.: Toward a Theory of Test Data Selection, *IEEE Trans. On Softw. Eng.*, 1975, pp. 156–173.
3. Srivastava, P.R.: Test Case Prioritization, Journal of Theoretical and Applied Information Technology, Vol. 4, No. 3, pp. 178–181.
4. Bryce, R.C., Colbourn, Ch.C.: Prioritized Interaction Testing for Pairwise Coverage with Seeding and Constraints, *Information and Software Technology*, 48, 2006, pp. 960–970.
5. Elbaum, S., Malishevsky, A., Rothermel, G.: Test Case Prioritization: A Family of Empirical Studies, *IEEE Transactions on Software Engineering*, 28(2), 2002, pp. 182–191.
6. Belli, F., Budnik, Ch. J., White, L.: Event-Based Modeling, Analysis and Testing of User Interactions — Approach and Case Study, *J. Software Testing, Verification & Reliability*, John Wiley & Sons, 16(1), 2006, pp. 3–32.
7. Belli, F., Budnik, C.J.: Test Minimization for Human-Computer Interaction, *J. Applied Intelligence*, 7(2), Springer, 2007, pp. 161–174.
8. Edmonds, J., Johnson, E.L.: Matching: Euler Tours and the Chinese Postman, *Math. Programming*, 1973, pp. 88–124.
9. Belli, F.: Finite-State Testing and analysis of Graphical User Interfaces, *Proc. 12th Int'l. Symp. Softw. Reliability Eng. (ISSRE'01)*, 2001, pp. 43–43.
10. Belli, F., Budnik, Ch. J., White, L.: Event-Based Modeling, Analysis and Testing of User Interactions — Approach and Case Study, *J. Software Testing, Verification & Reliability*, John Wiley & Sons, 16(1), 2006, pp. 3–32.
11. F. Belli, F., Budnik, C.J.: Test Minimization for Human-Computer Interaction, *J. Applied Intelligence*, 7(2), Springer, 2007, pp. 161–174.
12. Eminov, M.E.: Fuzzy c-Means Based Adaptive Neural Network Clustering. *Proc. TAINN-2003, Int. J. Computational Intelligence*, 2003, pp. 338–343.
13. Kim, D.J., Park, Y.W., Park, D.J.: A Novel Validity Index for Clusters, I*EICE Trans. Inf & System*, 2001, pp. 282–285.
14. Eminov, M., Gökçe, N.: Neural Network Clustering Using Competitive Learning Algorithm, *Proc. TAINN 2005*, 2005, pp. 161–168.
15. Maeda, M., Miyajima, H., Marashima, S.: An adaptive Learning and Self-Deleting Neural Network for Vector Quantization, *IEICE Trans. Fundamentals,*1996, pp. 1886–1893.
16. Maeda, M., Miyajim, H.: Competitive Learning Algorithm Founded on Adaptivity and Sensitivity Deletion Method, *IEICE Trans. Fundamentals*, 2000, pp. 2770–2774.

17. Belli, F., Budnik, Ch. J., Linschulte, M., Schieferdecker, I.: Testen Web-basierter Systeme mittels strukturierter, graphischer Modelle-Vergleich anhand einer Fallstudie, Model-based Testing 2006, LNI, Vol. P-94, pp. 266–273, GI, Bonn, October.
18. Fu, L.M.: Neural Networks in Computer Intelligence, McGraw-Hill, New York (1994).
19. Martinetz, T.M., Berkovich, S.G., Schulten, K.J.: "Neural-Gas" Network for Vector Quantization and Its Application to Times-series Prediction, *IEEE Trans. Neural Networks*, Vol. 4, No. 4, 1993, 558–569.
20. Kohonen, T.: Self-organization and Associative Memory. Springer-Verlag, Berlin, 1989.
21. Brownlee, J.: Lazy and Competitive Learning, Technical Report, Australia, 2007
22. George, F., Luger, William, A.: Stubblefield. Artificial Intelligence: Structures and Strategies for Complex Problem Solving, USA: Addison Wesley Longman Inc, 1997.
23. Kohonen, T.: The self-organizing map, *Proceedings of the IEEE*, Vol. 78, Sep, 1990, pp. 1464–1480.
24. Teuvo Kohonen, *Self-Organizing Maps*, Berlin Heidelberg:Springer-Verlag, 2001.
25. B. Fritzke, "Some competitive learning methods," Systems Biophysics, Institute for Neural Computation, Ruhr-Universitat Bochum, Germany, Apr 1997.
26. Rummelhart, D.E., Zipser, D.: Competitive Learning, *J. Cognitive Science*, 1985, pp. 75–112.
27. Graf, A.B.A., Smola, A.J., Borer, S.: Classification in a Normalized Feature Space using Support Vector Machines, *IEEE Trans. Neural Networks*, Vol. 14, No. 3, 2003, 597–605.
28. Gökçe, N., Eminov, M., Belli, F.: Coverage-Based, Prioritized testing Using Neural Network Clustering, The 21st International Symposium on Computer and Information Sciences, ISCIS 2006 Proceedings, LNCS volume 4263, pp. 1060–1071.
29. Belli, F., Eminov, M., Gökçe, N.: Prioritizing Coverage-Oriented Testing Process- An Adaptive-Learning-Based Approach and Case Study, The Fourth IEEE International Workshop on Software Cybernetics, IWSC2007, 24 July, Beijing, China.
30. Neate, B., Warwick, I., Churcher, N.: CodeRank: A New Family of Software Metrics, *Proc. Australian Software Engineering Conference – ASWEC 2006*, IEEE Comp. Press, 2006, pp. 369–377.

Chapter 2

STATISTICAL EVALUATION METHODS FOR V&V
OF NEURO-ADAPTIVE SYSTEMS

Y. LIU

Motorola Mobility Inc., Libertyville, IL 60048, USA
u00046@motorola.com

J. SCHUMANN

SGT/NASA Ames, Moffett Field, CA 94035, USA
Johann.M.Schumann@nasa.gov

B. CUKIC

Lane Department of Computer Science and Electrical Engineering
West Virginia University
Morgantown, WV 26505, USA
cukic@csee.wvu.edu

Biologically inspired soft computing paradigms such as neural networks are popular learning models adopted in online adaptive systems for their ability to cope with the demands of a changing environment. However, the acceptance of adaptive controllers is limited by the fact that methods and tools for the analysis and verification of such systems are still in their infancy. Generic Verification and Validation (V&V) procedures do not exist. The reliability of learning, performance, convergence or prediction for neural network models and their applications is hard to guarantee.

In this paper, we present several statistical evaluation methods proposed for the V&V of neuro-adaptive systems. These methods include support vector data description based novelty detection, statistical inference based evaluation of learning, and probabilistic measures of prediction performance. These evaluation methods are illustrated as dynamic monitoring tools for two types of neural networks used in the NASA Intelligent Flight Control System (IFCS) as adaptive learners: the Dynamic Cell Structure (DCS) network and the Sigma-Pi network.

1. Introduction

In recent years, the use of biologically inspired soft computing models such as neural networks for online adaptation to accommodate system failures and recuperate against environmental changes has revolutionized the operation of real-time automation and control applications. A variety of approaches for adaptive control, based upon self-learning computational models such as neural networks and fuzzy logic, have been developed.[1,2] The neural network models in these systems play an essential role as they adapt to the changes and provide a mechanism for accommodation of system failures. In the case of flight control applications, failure conditions require prompt reactions. The neural network is expected to adapt to such failure conditions fast enough to provide accommodation in real-time. However, the neural network's learning behavior is subject to the training data set acquired in flight, i.e., unknown at the system design phase. Unreliable predictions are likely to occur at poorly fitted regions. It is possible that abrupt environmental changes or unforeseen failure conditions beyond the learned domain will cause poor prediction performance. Such conditions challenge the use of neural network models in online adaptive systems and pose a serious problem for system verification and validation.

Applications such as flight control, process control, or robotic vehicles require adaptation because of the changes in the environment. For such systems, it is impossible to build a static model that is able to provide reliable performance in all situations, especially under unforeseen conditions. The popular approach to employ such systems is to develop an offline model, usually a pre-trained model for known functional regions. Then an online adaptive model is constructed in order to enable and optimize system's functionality in unknown functional regions. As an emerging application of online adaptive systems, adaptive flight control is one of the most promising real-time automation and control applications. The system achieves adaptability through judicious online learning, aids the adaptive controller to recover from operational damage (sensor/actuator failure, changed aircraft dynamics: broken aileron or stabilator, etc.). Some of these conditions are severe enough to be considered failure mode conditions that significantly affect system performance. National Aeronautics and Space Administration (NASA) conducted series of experiments evaluating adaptive computational paradigms (neural networks, AI planners) for providing failure accommodation capabilities in flight control systems following sensor and/or actuator failures. Experimental success suggests significant potential for further development and deployment of adaptive controllers.[3,4] Nevertheless, the (in)ability to provide a theoretically sound and practical verification and validation method remains one of the critical factors limiting wider use of *"intelligent"* flight control.[5-7]

Because adaptive systems include complex learning algorithms, no standardized way of performing performance analysis and V&V exists. Furthermore, certification authorities are reluctant to certify novel components, architectures, and software algorithms. The related work on verification and validation of neuro-adaptive system has not caught up with the advances of the learning techniques for adaptation. Existing approaches still heavily focus on static analysis using stability theories and empirical validation by vigorous offline testing. There is limited effort that attempts to provide a guidance to the V&V of neuro-adaptive systems. In 2002, NASA developed a software verification process guide[7] addressing the V&V issues of adaptive flight control systems. In addition, there is an increasing attention on statistical evaluation methods that basically monitor the online adaptation of the system in order to provide performance assessment. The use of Bayesian techniques to estimate neural network "quality" is presented in detail in Ref. 8. Another metric called validity index for Radial Basis Function (RBF) neural networks has been introduced in Ref. 9. Monitoring approaches for neuro-adaptive controllers, based on Lyapunov stability are discussed in Ref. 10.

Neural networks are widely used for function approximation, prediction and pattern recognition. The requirements on such models are usually described as satisfying certain criteria of precision and/or accuracy. Typical metrics used for performance evaluation of neural networks are Mean Square Error (MSE), Squared Error, etc. They are used to measure the learning performance of a neural network model. For prediction performance evaluation, the most popular metrics are prediction/confidence intervals defined to measure the reliability of network output. In the context of an neural network based adaptive control system, the online neural network is expected to promptly respond to (adapt to) environmental changes. Therefore, within a real time adaptive system, assuring the performance of the online neural network requires online evaluation of its adaptation performance. The evaluation should be performed to examine: (1) how fast the neural network responds to the changes, and (2) how well it accommodates the changes.

In addition to the nonlinearity and complexity of the neural network learning, system uncertainties coupled with real-time constraints make traditional V&V techniques insufficient for neuro-adaptive systems. Development and implementation of a non-conventional V&V technique

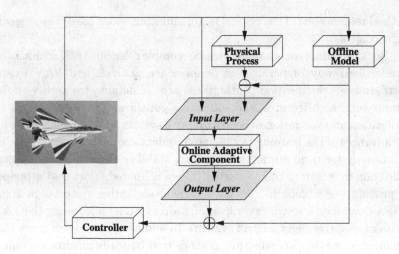

Fig. 1.   A typical adaptive flight control application.

is a challenging task. After a thorough examination of discrete formal analysis on several learning paradigms,[11] we understood the impact that environmental changes (learning data) have on system behavior. For a safety-critical system such as flight control, these changes must be observed, detected and well-understood before system deployment. Further, the adaptation caused by such changes must be monitored in real-time and the consequences of the adaptation must be measured and controlled.

Figure 1 illustrates a typical neuro-adaptive flight control application where the online adaptation is carried out by neural network models. Our research focuses on the learning performance of the neural network models as well as the two layers of the online adaptive component, referred to as the input layer and the output layer in Fig. 1. At the input layer, environmental changes are captured through sensor readings. These independent values are coupled with discrepancies between the response of a reference model and the output of an parameter identification model as they react to the current environmental conditions. Together, they are fed into the adaptive component as learning data. The learner adapts to the training data and may accommodate certain failure conditions. It is possible that abrupt/abnormal environmental changes, especially severe failure conditions, cause dramatic adaptation behavior and transient unreliable performance of the learner. As the adaptive element learns and adapts to the environmental changes, it also produces certain parameter values as corrections in response to the changes. Thus, at the output layer, the learner recalls what it has learned and generates derivative corrections as compensation to the offline model output. The corrected output is then sent to the next component (e.g., a controller) for further actions. The validity and correctness of such corrections are crucial to system safety.

At the input layer of a neuro-adaptive system, the failure-prone data need to be analyzed and potential failures detected. When a failure occurs, the learner learns the failure conditions and adapts to the corresponding environmental changes. At the output layer, the system safety relies greatly on the accommodation performance of the learner. The reliability of the predictions made by the learner depends on how well and how fast the learner adapts to failure conditions. In order to evaluate the accommodation performance of the learner and validate its prediction performance, a performance index such as confidence level must be measured as an indication of trustworthiness and reliability of the system output.

We seek for validation methods at all layers. Our investigation suggests a need for three different techniques: (1) near real-time failure detection at the system input layer, (2) online monitoring and evaluation of the neural network's learning performance, and (3) a feasible validation approach to validate the prediction performance at the system output layer. Thus, we propose a novel validation approach that consists of three statistical evaluation techniques to perform validation. The failure detection at the input layer relies on robust novelty detection techniques. Such techniques are used to examine the learning data on which the adaptive component is trained and detect the failures efficiently and accurately. The online monitoring estimates the adaptation performance of the neural network by analyzing the network's structural and parametric properties. The validation at the output layer is performed after the adaptation to verify the correctness of the output and evaluate the accommodation abilities of the learner.

Accordingly, we developed three statistical evaluation methods as a part the effort to validate the NASA Intelligent Flight Control System (IFCS). At the input layer, a real-time failure detection technique is implemented before the data can propagate into the learner and cause adaptation. As an independent novelty detector, it detects failures based on the knowledge collected from the nominal system behavior only. During learning, we estimate how the quality of each parameter of the network (e.g., weight) influences the output of the network by calculating parameter sensitivity and parameters confidence for the neural network. When failure occurs, the online neural network accommodates the failure in real-time

and predicts certain parameter values as compensations/corrections to
the offline model output in response to the failure conditions. Hence,
at the output layer, an online reliability measure associated with each
prediction produced by the learner is generated for validity check. At
last, all approaches serve as major components of a validation framework
that can be generalized to extensive online neural network-based adaptive
control systems.

In the rest of this paper, Section 2 presents our findings in related
research in V&V of neuro-adaptive systems as well as the existing
approaches for validating neural networks. Section 3 first introduces a
neuro-adaptive flight control application where two different types of neural
networks are deployed as its online learning component for adaptation
for failure accommodation. Then three different statistical approaches are
described: a failure detection method using a fast support vector data
description algorithm, the sensitivity analysis for both neural networks to
dynamically monitor the networks' learning performance; and the validity
index (VI) as a reliability measure for the network output of Dynamic Cell
Structures (DCS) and confidence levels for the output of Sigma-Pi network.
Section 4 concludes the paper with a few remarks on how these statistical
methods can be applied to other neuro-adaptive systems.

2. V&V of Neuro-Adaptive Systems

As an emerging paradigm, the neuro-adaptive systems are becoming more
popular. Yet, the research efforts on V&V of such systems are still rare.
Recently, much effort has been dedicated to analyzing theoretical properties
of online learning. Methods that can be applied in an online fashion to
assure the performance are the subject of interest too. Most proposed
approaches address the significant impact of learning data on system
behavior and investigate certain properties of learning in order to ensure
system safety through online monitoring or rule checking. We summarize
the existing approaches into two categories, namely, static V&V approaches
and dynamic V&V approaches.

2.1. *Static V&V Approaches*

Most static verification and validation methods focus on the inherent
properties of online adaptive learning. These methods are used to
theoretically establish the correctness of the learning behavior with

respect to the requirement specifications. Researchers often employ mathematically rigorous theories to prove functional properties and/or operational properties. Approximation theory is used to prove certain families of neural networks as universal approximators. Examples are Multi-Layer Perceptron (MLP) networks and RBF networks. These two types of neural networks are very popular choices for adaptive learners and have been proven to be universal approximators.[12] It is claimed that an MLP (or RBF) network with sufficiently large number of neurons can approximate any real multi-variate continuous function on a compact set. However, the number of neurons has to be pre-defined before the system deployment. Usually, the number of neurons an MLP/RBF network requires to map a complex function may have to be very large. In the instance of an online adaptive system, the proven theory offers little guidance in validating the online learning performance of neural network based adaptation.

Empirical methods are also available for testing and verifying the adaptive learner against certain safety and reliability requirements.[13] The train-test-retrain scheme for validating neural network performance is a popular V&V approach. Yet, this time-consuming procedure is not suitable for an online adaptive system due to the fact that the network has to learn in near real-time. The bias-variance analysis provides guidelines on the generalization performance of a learning model, but can hardly be applied to improve the prediction performance of an online learning system.

The static verification methods using formal methods and approximation theories provide an insightful analysis on neural network-based adaptive systems. For most neural computing systems, empirical methods are practical for performance validation. However, there is a widespread agreement that such static approaches are inapplicable to online adaptive systems, whose function evolves over time and responds differently to various environmental changes.

2.2. *Dynamic V&V Approaches*

Instead of statically validating the learning properties of a neuro-adaptive system, dynamic approaches adopt the online monitoring approach to cope with the evolving performance of neural networks. These methods concentrate on two different aspects (phases).

1. For any learning system, training data is always gathered before the learner is used for prediction. Verification of the training data includes

the analysis of its appropriateness and comprehensiveness. The strong emphasis on domain specific knowledge, its formal representation and mathematical analysis is suggested in Ref. 14. Del Gobbo and Cukic propose the analysis of the neural network with respect to conditions implying the existence of the solution (for function approximation) and the reachability of the solution from any possible initial state. Their third condition can be interpreted as condition for preservation of the learned information. This step is not fully applicable to on-line learning applications since training data are related to the real-time evolution of the system state, rather than the design choice. However, as proven by our previous investigation using formal methods,[11] the training data has a very significant impact on system behavior. In a safety-critical system, the ability of "novelty detection" is crucial to system safety. It helps to detect suspicious learning data that is potentially hazardous to the system operation.

2. Online monitoring techniques have been proposed to validate the learning process. In a recent survey of methods for validating online learning neural networks, Raz[13] acknowledges the online monitoring techniques as a significant potential tool for the future use. Another promising research direction, according to Raz, is periodic rule extraction from an online neural network and partial (incremental) re-verification of these rules using symbolic model checking. In Ref. 15, Taylor *et al.* focus their effort on the Dynamic Cell Structure. They propose a prototype for real-time rule extraction in order to verify the correctness of DCS learning performance. In Ref. 16, M. Darrah *et al.* present rule extraction from DCS network learning and suggest future examination of performance based on such rules. Practical hurdles associated with this approach include determining the frequency of rule extraction and impracticality of near real-time model checking of complex systems.

Yerramalla *et al.* develop a monitoring technique for the DCS neural network embedded in the IFCS[17,18] based on Lyapunov stability theory. The online monitors operate in parallel to the neural network with the goal of determining whether (or not), under given conditions, the neural network is convergent, meaning that all state transition trajectories converge to a stationary state. The online monitor is theoretically founded and supported by an investigation of mathematical stability proofs that can define the engagement (or disengagement) of the online monitor.

We notice that efforts exist to validate the prediction performance, where the system is in operation after learning for a certain period of time. In some cases, neural networks are modified to provide support for testing based (or online) validation of prediction performance. For example, Leonard *et al.*[9] suggest a new architecture called Validity Index Net. A Validity Index network is a derivative of Radial Basis Function (RBF) network with the additional ability to calculate confidence intervals for its predictions based on the probability density of the "similar" training data observed in the past.

2.3. V&V of Neural Networks

Because online learning systems are often used in life-critical (e.g., flight control) and mission-critical (e.g., space) applications, they are subject to strict certification standards, leaving a wide technological gap between the requirements of the application domain and the capabilities of available technologies; the goal of our research is to narrow this gap. Hence, we survey existing approaches to V&V of neural networks.

Traditional literature describes adaptive computational paradigms, neural networks in particular, with respect to their use, as function approximators or data classification tools. Validation on these systems is usually based on a train-test-re-train empirical procedure. Some bibliographic references also propose methods as part of the training algorithm of neural networks for validation.[5,19] The ability of interpolating and/or extrapolating between known function values is measured by certain parameters through testing. This evaluation paradigm can be reasonably effective only for pre-trained adaptive systems, which does not require online learning and adaptation and remain unchanged in use. In Ref. 20, Fu interprets the verification of a neural network to refer to its correctness and interprets the validation to refer to its accuracy and efficiency. He establishes correctness by analyzing the process of designing the neural network, rather than the functional properties of the final product. Gerald Peterson presents another similar approach in Ref. 21 by discussing the software development process of a neural network. He describes the opportunities for verification and validation of neural networks in terms of the activities in their development life cycle.

Verification of the training process typically examines the convergence properties of the learning algorithm, which is usually pre-defined by some criteria of error measure. In Ref. 22, Hunt *et al.* investigate all different methods for error estimation techniques and make detail comparison among them. Nonetheless, effective evaluation methods of interpolation and extrapolation capabilities of the network and domain specific verification activities are still based on empirical testing.[23] Literature addressing the problem analytically are rare.

Among existing approaches of V&V of dynamic neural networks, statistical evaluation methods are regarded as practical and effective in detecting novelties and validating learning and prediction performances. In an attempt to solve the dilemma of plasticity and stability for neural networks, Grossberg[24,25] derives a new paradigm, referred to as the Adaptive Resonance Theory (ART-1/2/3). Within such a network, there are two components charging seen and unseen data respectively. As interesting as is, it provides better understanding for our problem other than applicable tools for validation and verification. Another new architecture which can be extended for our research goal is the aforementioned Validity Index network presented by Leonard *et al.*[9] The validity index in a Radial Basis Function neural network is a confidence interval associated with each network prediction for a given input. It can be viewed as a reliability measure for the RBF network's prediction performance.

3. Statistical Evaluation of Neuro-Adaptive Systems

Based on the investigation results presented in the previous section, it is our conclusion that dynamic and efficient statistical evaluation methods can support effective and practical monitoring techniques for V&V of neuro-adaptive systems. In this chapter, we describe three novel methods that target different aspects of the neural network based adaptation performance evaluation, all of which are deployed and tested in a case study for V&V of the IFCS.

3.1. *Neural Network-Based Flight Control*

We illustrate our approach with the NASA F-15 IFCS project. Its aim is to develop and test-fly a neuro-adaptive intelligent flight control system for a manned F-15 aircraft. Two principal architectures have been developed: the Gen-I architecture uses a DCS neural network as its online adaptive component, the Gen-II architecture a Sigma-Pi network. Both network architectures and their training algorithms will be described in more details below. The target aircraft for this controller is a specifically configured NASA F-15 jet aircraft, which has been highly modified from a standard F-15 configuration to include canard control surfaces, thrust vectoring nozzles, and a quad-redundant digital fly-by-wire flight control system. As visible in Fig. 1, the canards, which are small winglets, are located in front of the wings. By moving them, the airflow over the wing can be modified in a wide range. Thus, this aircraft can be used to simulate failures like damage to the wings during test flights.

Figure 2 shows the basic architecture of the Gen-I and Gen-II controllers: pilot stick commands Θ_{cmd} are mixed with the current sensor readings Θ (e.g., airspeed, angle of attack, altitude) to form the desired behavior of the aircraft (measured as roll-rate, pitch-rate, and yaw-rate). From these data, the PID controller calculates the necessary movements of the control surfaces (e.g., rudder, ailerons) and commands the actuators. The controller incorporates a model of the nominal aircraft dynamics. If the aerodynamics of the aircraft changes (e.g., due to a damaged wing or a stuck rudder), there is a deviation between desired and actual state. The neural network is trained during operation to minimize this deviation. Whereas in the Gen-I architecture, the appropriate control derivatives are modified with a neural network, Gen-II uses a dynamic inverse controller with control augmentation, i.e., the neural network produces a control correction signal. The inputs of the neural network are typically the current state of the aircraft (i.e., the sensor signals), the commanded input, and the correction signal of the previous time frame. For details on the control architecture see Refs. 1, 26.

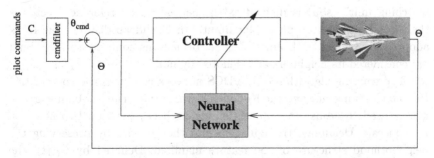

Fig. 2.   IFCS Generic Adaptive Control Architecture.

3.2. *The Neural Networks*

For safety-critical systems, a "black box" approach to the neural network's performance assessment is definitely insufficient. We therefore must estimate how the quality of each parameter of the network (e.g., weight) influences the output of the network by calculating parameter sensitivity and parameters confidence for the neural networks implemented in both Gen-I and Gen-II IFCS. Such calculations are very specific to the inherent structural and learning properties of the neural network models. Thus, we first introduce these two different types of neural networks as follows.

3.2.1. *Dynamic Cell Structure Network*

The Dynamic Cell Structures network is derived as a dynamically growing structure in order to achieve better adaptability. DCS can be seen as a special case of Self-Organizing Map (SOM) structures as introduced by Kohonen[27] and further improved to offer topology-preserving adaptive learning capabilities that can respond and learn to abstract from a much wider variety of complex data manifolds.[28,29] In the IFCS Gen-I controller, the DCS provides derivative corrections as control adjustments during system operation. It has been proven to outperform Radial Basis Function (RBF) and Multi-Layer Perceptron network models.[15,30] As a crucial component of a safety critical system, DCS network is expected to give robust and reliable prediction performance in operational domains.

The DCS network adopts the self-organizing structure and dynamically evolves with respect to the learning data. It approximates the function that maps the input to the output space. At last, the input space is divided into different regions, referred to as the Voronoi regions.[28,29,31] Each Voronoi region is represented by its centroid, a neuron associated with its reference vector known as the "best matching unit" (bmu). Further, a "second best matching unit" (sbu) is defined as the neuron whose reference vector is the second closest to a particular input. An Euclidean distance metric is adopted for finding both units. The set of neurons connected to the bmu are considered its neighbors and denoted by nbr.

The training algorithm of the DCS network combines the competitive Hebbian learning rule and the Kohonen learning rule. The Hebbian learning rule is used to adjust the connection strength C_{ij} between two neurons. It induces a Delaunay Triangulation into the network by preserving the neighborhood structure of the feature manifold. Denoted by $C_{ij}(t)$, the connection between neuron i and neuron j at time t is updated as follows:

$$
C_{ij}(t+1) = \begin{cases}
1 & (i = bmu) \land (j = sbu) \\
0 & (i = bmu) \land (C_{ij} < \theta) \\
 & \land (j \in nbr \backslash \{sbu\}) \\
\alpha C_{ij}(t) & (i = bmu) \land (C_{ij} \geq \theta) \\
 & \land (j \in nbr \backslash \{sbu\}) \\
C_{ij}(t) & i, j \neq bmu
\end{cases}
$$

where α is a pre-defined forgetting constant and θ is a threshold preset for dropping connections.

The Kohonen learning rule is used to adjust the weight representations of the neurons $(\vec{w}_i)$, which are activated based on the best-matching methods during the learning. If needed, new neurons are inserted. After learning, when DCS is used for prediction (the recall mode), it will recall parameter values at any chosen dimension. It should be noted that the computation of an output is different from that during training. When DCS is in recall mode, the output is computed based on two neurons for a particular input. One is the bmu of the input; the other is the closest neighbor of the bmu other than the sbu of the input. In the absence f neighboring neurons of the bmu, the output value is calculated using the bmu only. Since our performance estimation does not depend on the specific learning algorithm, it will not be discussed in this paper. For details on DCS and the learning algorithm see Refs. 28, 29, 31, 32.

Over every training cycle, let $\Delta w_i = w_i(t+1) - w_i(t)$ represent the adjustment of the reference vector needed for neuron i, the Kohonen learning rule followed in DCS computes $\Delta \vec{w}_i$ as follows.

$$
\Delta \vec{w}_i = \begin{cases}
\varepsilon_{bmu}(m - w_i(t)) & i = bmu \\
\varepsilon_{nbr}(m - w_i(t)) & i \in nbr \\
0 & (i \neq bmu) \land (i \notin nbr)
\end{cases}
$$

where $\vec{m}$ is the desired output, and $0 < \varepsilon_{bmu}, \varepsilon_{nbr} < 1$ are predefined constants known as the learning rates that define the momentum of the update process. For every particular input, the DCS learning algorithm applies the competitive Hebbian rule before any other adjustment to ensure that the sbu is a member of nbr for further structural updates.

The DCS learning algorithm is briefly described in Fig. 3. According to the algorithm, N is the number of training examples. Resource values are computed at each epoch as local error measurements associated with each neuron. They are used to determine the sum of squared error of the

```
Initialization;

Repeat until stopping criterion is satisfied;

{

        Repeat N times

        {

                Determine the bmu and sbu;

                Update lateral connections;

                Adjust the weights;

                Update resource values;

        }

        If needed, a new neuron is inserted;

        Decrement resource values;

}
```

Fig. 3. A brief description of the DCS learning algorithm.

whole network. Starting initially from two connected neurons randomly selected from the training set, the DCS learning continues adjusting its topologically representative structure until the stopping criterion is met. The adaptation of lateral connections and weights of neurons are updated by the aforementioned Hebbian learning rule and Kohonen learning rule respectively. The resource values of the neurons are updated using the quantization vector. In the final step of an iteration, the local error is reduced by inserting new neuron(s) in certain area(s) of the input space where the errors are large. The whole neural network is constructed in a dynamic way such that in the end of each learning epoch, the insertion or pruning of a neuron is triggered when necessary.

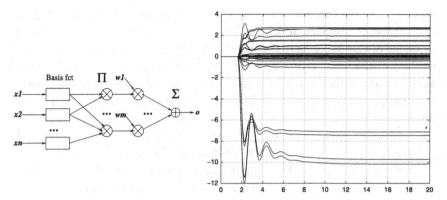

Fig. 4. Architecture of $\Sigma\Pi$ network (left). Development of the NN weights over time during adaptation (right). The failure occurred at $t = 1.5\,\mathrm{s}$.

3.2.2. *Sigma-Pi Neural Network*

The IFCS Gen-II controller uses a Sigma-Pi ($\Sigma\Pi$) neural network,[33] where the inputs x are subjected to arbitrary basis functions (e.g., square, scaling, logistic function). Then Cartesian products of these function values are calculated. The final output of the network o is a weighted sum of these products (Fig. 4 (left)). The functionality of a Sigma-Pi network is defined by

$$o = \sum_i w_i \prod_j \beta(x_j)$$

with weights w_i and $\beta(x_j)$ the basis functions. The sequence of the operators ($\Sigma\Pi$) gave this network architecture its name. During the training, the weights w_i are modified as to minimize the tracking error of the controller. The neural network is trained in an online fashion using the e-modification rule,[1] an improved variant of the gradient descent learning rule. As our approach to confidence and sensitivity analysis does not depend on the specific training algorithm for the network, it will not be discussed here.

Figure 4 (right) shows how the network weights w_i develop over time during an operational scenario. At $t = 1.5\,\mathrm{s}$, a failure occurs, triggering adaptation of the neural network.

3.3. *Failure Detection Using Support Vector Data Description*

In a safety-critical system like IFCS, a novelty detector can provide failure detection capabilities based on the the duration and the degree of the novel event. We can also use the novelty detector to observe and help understand the impact of the failure conditions on the adaptive learning behavior of the online neural network. An effective and efficient detection method at the input layer offers not only reliable inference on the learning data, but also sufficient response time for manual intervention to take place. For example, when an extremely severe failure occurs, we might have to discard the learning data and prevent it from impairing the performance of the learner. Prompt human operations is required. Efficient evaluation and detection will give the system the freedom to take necessary actions.

Support Vector Data Description (SVDD) has a proven record as a one-class classification tool for novelty detection.[34,35] In particular, it is highly advantageous in applications whose "abnormal" event/data is extremely costly or near impossible to obtain. We apply SVDD to realize the novelty detection as a major step of our validation approach. However, due to space complexity of matrix operations, the optimization process becomes memory and time consuming when n, the size of training set increases. Hence, efficiency needs to be improved for data sets of large size. We present an algorithm that first reduces the space complexity by breaking the training data set into subsets at random and apply SVDD to each subset. Then, based on two lemmas of random sampling and SVDD combining, we merge the data descriptions into a "common decision boundary". Provided with the fact that usually the number of support vectors is relatively few with respect to n, the search for a common decision boundary for a training data set of size n in a d-dimension space can be bounded at $O(d^{\frac{3}{2}} n^{\frac{3}{2}} \log n)$ steps.

The method of support vector data description originates from the idea of finding a sphere with the minimal volume to contain all data.[36] Given a data set S consisting of N examples $x_i, i = 1, \ldots, N$, the SVDD's task is to minimize an error function containing the volume of the sphere. With the constraint that all data points must be within the sphere, which is defined by its radius R and its center a, the objective function can be translated into the following form by applying Lagrangian multipliers:

$$L(R, a, \alpha_i) = R^2 - \sum_i \alpha_i \{R^2 - (x_i - a)^2\},$$

where $\alpha_i > 0$ is the Lagrange multiplier. L is to be minimized with respect to R and a and maximized with respect to α_i. By solving the partial derivatives of L, we also have:

$$\sum_i \alpha_i = 1;$$

and

$$a = \sum_i \alpha_i x_i,$$

which gives the Lagrangian with respect to α_i:

$$L = \sum_i \alpha_i (x_i \cdot x_i) - \sum_{i,j} \alpha_i \alpha_j (x_i \cdot x_j),$$

where $\alpha_i \geq 0$ and $\sum_i \alpha_i = 1$. By replacing some kernel functions $K(x, y)$ with the product of (x, y) in the above equations, in particular, the Gaussian kernel function $K(x, y) = \exp(-\|x - y\|^2 / s^2)$, we have:

$$L = 1 - \sum_i \alpha_i^2 - \sum_{i \neq j} \alpha_i \alpha_j K(x_i, x_j).$$

By applying kernel functions, we can have a better description of the boundary. The application of kernel functions injects more flexibility to the data description. According to the solution that maximizes L, a large portion of α_i's become zero. Some α_i's are greater than zero and their corresponding objects are those called support objects. Support objects lie on the boundary that forms a sphere that contains the data. Hence, object z is accepted by the description (within the boundary of the sphere) when:

$$\|z - a\|^2 = \left(z - \sum_i \alpha_i x_i\right)\left(z - \sum_i \alpha_i x_i\right) \leq R^2.$$

Similarly, by applying the kernel function, the formula for checking an object z now becomes:

$$1 - 2\sum_i \alpha_i K(z, x_i) + \sum_{i,j} \alpha_i \alpha_j K(x_i, x_j) \leq R^2.$$

Since the SVDD is used as a one-class classifier, in practice, there is no actual outliers well defined other than those we randomly draw from the rest of the space outside the target class. Hence, by applying the SVDD, we can only obtain a relatively sound representation of the target class. To detect outliers, more precise criteria should be inferred from empirical testings or pre-defined thresholds. In addition, most real-world data are highly-nonlinear and thus a sphere-like boundary would be almost useless for novelty detection. In order to obtain a "soft boundary", Tax *et al.* introduces the parameter C, pre-defined as tradeoff between the volume of our data description and the errors. In general, $C \leq \frac{1}{nf}$, where f is the fraction of outliers that are allowed to fall outside the decision boundary over the total number of data points in S.[36] And the Lagrangian form L is rewritten as:

$$L(R, a, \xi) = R^2 + C \sum_i \xi_i.$$

And the constraints are:

$$\|x_i - a\|^2 \leq R^2 + \xi_i, \forall i.$$

By applying Lagrangian multipliers, the above Lagrangian can be simplified as follows.

We are to maximize L with respect to α:

$$L = \sum_i \alpha_i(x_i \cdot x_i) - \sum_{i,j} \alpha_i \alpha_j (x_i \cdot x_j)$$

with $0 \leq \alpha_i \leq C$ and $\sum_i \alpha_i = 1$.

There are three objects obtained from the final solution of maximizing L, which are outliers, support vectors and the rest of the data points that are confined by the boundary. Outliers are those data points that lie outside the boundary. They are considered "novelties" in this case. Support vectors are those data points that sit on the boundary. For support vectors, $0 < \alpha_i \leq C$. A large portion of the data have $\alpha_i = 0$ and these are the data points that lie inside the boundary. Therefore, the center of the hyper-sphere are in fact determined by a very small portion of the data, the support vectors. And the fraction of those data points that are support vectors is a *"leave-one-out estimate of the error on the target data set."*[36] Therefore, we have:

$$E[P(error\ on\ the\ target\ set)] = \frac{number\ of\ support\ vectors}{N}.$$

Furthermore, SVDD can also produces a "posterior probability-like" novelty measure for each testing data point that falls outside the boundary. Based on the assumption that the outliers are distributed uniformly in the feature space, Tax maps the distance from the outlier object to the defined decision boundary to a novelty measure. It is a quantified measure that indicates the degree of novelty of this particular object with respect to the target class. The mathematical definition of this mapping follows.

$$p(z|O) = \exp(-d(z|T)/s)$$

where $p(z|T)$ is the probability that z belongs to the outlier class; $d(z|T)$ is the distance from object z to the decision boundary in the feature space and s is the kernel width.

By applying the SVDD method, we can obtain a sound representation of the target class. To detect outliers (in our case, system failure conditions), a precise criterion should be inferred from empirical testing or pre-defined thresholds. The greater the the distance from the bounded region, the rougher the boundary. Therefore, the sensitivity of outlier detection may be changed. In practice, a pre-defined threshold can be used as the maximum distance of a data point from the center, which the system can tolerate. Such pre-defined thresholds need sufficient testing within each specific data domain.

SVDD has been successfully applied to different domains[34,35,37] and has many appealing features as a very promising novelty detection tool. However, SVDD faces the challenge of space and time complexity when the data size reaches a certain number and the matrix operation becomes extremely time-consuming. In reality, the control system of FCS runs at the speed of 20 Hz and generates 200 data points within 10 seconds. A large amount of data is needed to obtain a sound and meaningful data description for the nominal regions. Hence, we must improve the efficiency of SVDD to be used as a potential tool for novelty detection to realize a major step of our validation approach.

Based on two lemmas, the random sampling lemma and the combining lemma,[32] we develop a fast SVDD algorithm as shown in Fig. 5. The improved SVDD algorithm consists of two major steps, i.e., decomposition and combination. As an interesting application of the simple random sampling lemma and the combination lemma, the proposed fast SVDD algorithm demonstrates improvement of efficiency with respect to run-time complexity and memory utilization. In the context of previously proposed V&V approach for online adaptive systems, the fast SVDD algorithm can be applied to defining boundaries for "nominal" regions and used for real-time failure detections.

We notice that the improvement of the algorithm relies on the fact that by choosing $m \approx \sqrt{dn}$, the number of violators is bounded at m and the algorithm will converge to the global solution in a finite number of iterations. In the worst case scenario where the entire data set has to be learned, the number of iterations of our algorithm is $\sqrt{\frac{n}{d}}$, and for each step of decomposition and combination, the common decision boundary is found at a cost of $O(d^{\frac{3}{2}}n^{\frac{3}{2}})$. Therefore, the expected running complexity in total is

1. $W_1 \leftarrow$ **arbitrarily select m points from S;**
   Find the solution to W_1, denote it by P_1;
   $X_1 \leftarrow$ **the support vectors of (W_1, P_1), $x_1 \leftarrow |X_1|$;**
   $W_2 \leftarrow$ **select m points from $S \setminus W_1$;**

2. **Find the solution to W_2, denote it by P_2 ;**
   $X_2 \leftarrow$ **the support vectors of (W_2, P_2), $x_2 \leftarrow |X_2|$;**

3. $V_1 \leftarrow$ **the violators of (W_1, P_1), $v_1 \leftarrow |V_1|$;**
   if $v_1 = 0$, return (S, P_1) as the final solution.
   $V_2 \leftarrow$ **the violators of (W_2, P_2), $v_2 \leftarrow |V_2|$;**
   if $v_2 = 0$, return (S, P_2) as the final solution.

4. $R \leftarrow (X_1 \cup X_2) \cup ((V_1 \cup V_2) \cap \overline{W_1 \cup W_2})$;
   if $|R| < m$, add $m - |R|$ points from $W_1 \cup W_2$ into R;
   find the solution to R, denoted by Q,
   $X \leftarrow$ **support vectors of (R, Q);**

5. $V \leftarrow$ **the violators of (R, Q) in $S \setminus (W_1 \cup W_2)$;**
   if $|V| = 0$
       return (S, Q) as the final solution.
   if $|V| > m$
       $W_2 \leftarrow$ **randomly sample m points from V;**
   else
       $W_2 \leftarrow V \cup$ **{randomly sampled $m - |V|$ pts from $S \setminus V$ };**
   end

6. $W_1 \leftarrow X \cup$ **{ randomly sampled $m - |X|$ points from all learned non-SV, non-violator points };**

7. **Repeat $2 - 6$ until a global solution is found.**

Fig. 5.   A decomposing and combining algorithm for SVDD.

$O(dn^2)$. However, we can expect the iteration to be ended within $\log n$ steps and thus we can reach a running time of $O(d^{\frac{3}{2}} n^{\frac{3}{2}} \log n)$. On the other hand, since the size of the problem can be reduced into a level that computers can handle it at a reasonable speed, the memory utilization becomes efficient and less error-prone. Yet, m is not strictly fixed at $\sqrt{dn}$. Instead, it can be any number that is larger than the expected number of support vectors but near "optimally small" in terms of system specifications. The convergence will be reached in a finite number of steps due to the fact the objective

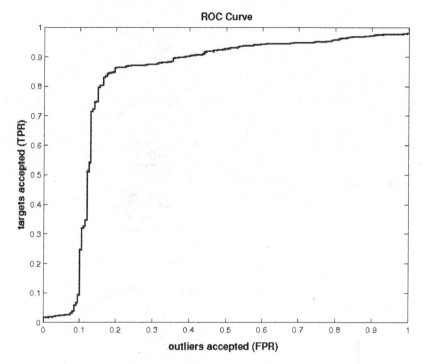

Fig. 6.   The ROC Curve of the SVDD tool.

function is convex and quadratic. However, the analysis of complexity in these cases might be slightly different. We first simulate one run of nominal flight conditions of 40 seconds with a segment of 800 data points saved. After running the fast SVDD on the nominal data, we obtain a sound data description of nominal flight conditions. A representative ROC curve is given in Fig. 6. By varying the value of the classification threshold we can obtain the differing SVDD classification characteristics in terms of combining false negatives and false positives. Based on the ROC curve and in line with system requirements, the specific operating point we choose for our SVDD tool is to allow 15% of nominal data classified as outliers.

We use the boundary formed by the proposed fast SVDD algorithm to test the failure mode simulation. Novelty detection results are shown in Fig. 7(b). Circles in Fig. 7(a) represent failure mode simulation data. The locked control surface failure results in input data points data falling outside the SVDD boundary. The novelty measures shown in Fig. 7(b) are probability-like measures computed for each data point based on the

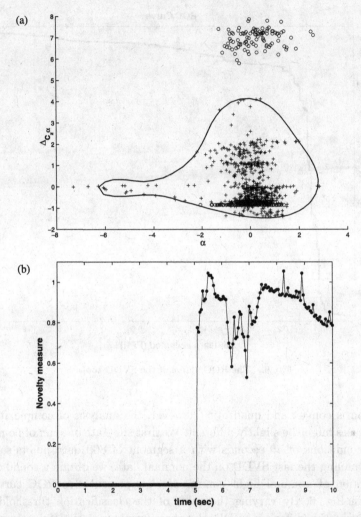

Fig. 7.   Novelty detection. (a): SVDD of nominal flight simulation data is used to detect novelties. (b): Novelty measures returned by SVDD tool for each testing data point.

distance from the SVDD boundary. In this plot, x-axis represents the time and y-axis represents the novelty measures calculated by the SVDD tool. We can see from the plot that, after 5 seconds when the failure occurs, SVDD detects the abnormal changes and returns high novelty measures. This demonstrates effective and accurate detection capabilities of our SVDD detector. For a full set of experimental results, please refer to Ref. 32.

3.4. *Evaluating Network's Learning Performance*

A good knowledge of the neural network's learning performance is important in order to obtain information under which a neuro-adaptive controller might exceed it robustness limits. Any feedback controller can always handle small deviations between the model and the actual plant dynamics. However, this robustness is strictly limited by the design of the controller, so large deviations between model and plant or very noisy signals can cause severe problems. A close examination of the internal structure and parameters usually reveals key information of the network's learning performance.

For the analysis of any controller's behavior, it is important to estimate its sensitivity with respect to input perturbations. A badly designed control system might amplify the perturbations, which could lead to oscillations and instability. The higher the *robustness* of the controller, the less influence arises from input perturbations. It is obvious that such a metric (i.e., $\frac{\partial \mathbf{o}}{\partial \mathbf{x}}$ for outputs $\mathbf{o}$ and inputs $\mathbf{x}$) is also applicable to an adaptive control system. For an adaptive component, like a neural network, the estimation of the sensitivity is a "black box" method, i.e., no knowledge about the internal structure or parameters is necessary.

In our approach, we focus on *parameter sensitivity*. This means, we calculate $\frac{\partial \mathbf{o}}{\partial p}$ for each of the adjustable parameters $p \in \mathcal{P}$. For a neural network, $\mathcal{P}$ is comprised of the network weights w_i, for the DCS network, it is the reference vectors of the neurons $\vec{w}_i$. During training of the network, these parameters are adjusted to minimize the error. Depending on the architecture of the adaptive controller, the network can be pre-trained, i.e., the parameters are determined during the design phase ("system identification"), or the parameters are changing while the system is in operation ("online adaptation"). In both cases, one needs to know, which influence the actual values of the parameters have on the output of the neural network: if the influence of a parameter or neuron is negligible, then this neuron might be removed from the network. On the other hand, extremely high sensitivity might cause numerical problems. Even more information can be obtained if we consider each parameter of the neural network not as a scalar value, but as a probability distribution. Then, we can formulate the sensitivity problem in a statistical way. The probability of the output $\mathbf{o}$ of the neural network is $p(\mathbf{o}|\mathcal{P}, \mathbf{x})$ given parameters $\mathcal{P}$ and inputs $\mathbf{x}$. If we again assume a Gaussian probability distribution, we can define our parameter confidence as the variance $\sigma_{\mathcal{P}}^2$. In contrast to calculating the network output confidence value, we do not marginalize over the weights, but over the inputs.

3.4.1. *A Sensitivity Metric for DCS Networks*

Within the IFCS Gen-I, the DCS networks are employed for online adaptation/learning. Their parameters (connection strength C_{ij} and reference vectors $\vec{w}_i$) are updated during system operation. It should be noted that the connection strength C does not contribute to the network predictions while it is in recall mode. This implies that the sensitivity of the connection strength is merely a structure related parameter that influences the reference vectors instead of the network output. We therefore only measure the sensitivity of the reference vector of the DCS network. Using the simulation data obtained from the IFCS Gen-I simulator, we calculate the parameter sensitivity s and its confidence σ^2 after each learning epoch during a flight scenario. The sensitivity analysis has been conducted on a N-dimension space, where N is the number of dimensions of the input space.

Figure 8 shows two sensitivity snapshots at different times of the simulation where the network has been trained with two-dimensional data. Each neuron is associated with a two-dimensional sensitivity ellipse. At the beginning of the simulation, the network is initialized with two neurons whose reference vectors represent two randomly selected training data points. The network continues learning and adjusts its own structure to adapt to the data. Figure 8 (left) shows the situation at $t = 5.0\,\mathrm{s}$. Figure 8 (right) shows the situation at $t = 10.0\,\mathrm{s}$. At $t = 5.0\,\mathrm{s}$, most neurons

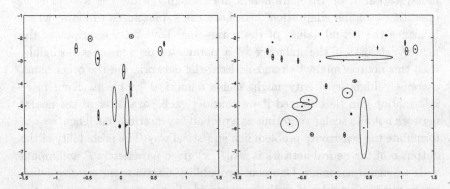

Fig. 8.   Sensitivity analysis for DCS networks.

exhibit relatively large sensitivity, while only a few (31%) neurons have small sensitivity values. However, at $t = 10.0$ s, when the network has well adapted to the data, Fig. 8 (right) clearly indicates that now most (78%) neurons have small sensitivity values.

3.4.2. *A Sensitivity Metric for Sigma-Pi Networks*

We also implement the sensitivity analysis for the online adaptive Sigma-Pi network of the IFCS Gen-II controller. We calculate the parameter sensitivity s and its confidence σ^2 for the network parameters w_i at each point in time during a flight scenario. Figure 9 shows two sensitivity snapshots at various stages of the scenario (roll axis shown). This network consists of 60 nodes and 60 weights. At the beginning of the scenario, all parameters of the network are set to zero, giving (trivially) in the same sensitivity. At $t = 1.5$, a failure is induced into the system. In order to compensate for the failure, the network weights adapt (see Fig. 4 (right)). Figure 9 (left) shows the situation at $t = 5.0$ s. A considerable amount of adaptation and weight changes has taken place already. However, the confidence for each of the 60 neurons is still relatively small, as indicated by the large error bars.

After approximately 20 seconds, the neural network is fully trained. Figure 9 (right) now shows quite different values for the sensitivity. Whereas the sensitivity for most of the neurons is really small now, a few (here 7) neurons exhibit high sensitivity. Although their σ^2 is somewhat larger than that for the other neurons, a clear distinction between the different groups can be made. The weights with low sensitivity are obviously candidates for pruning the network or improvement of the learning rule.

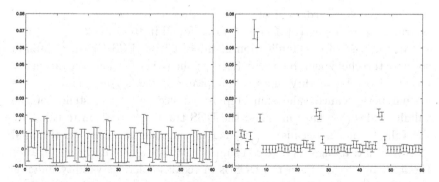

Fig. 9. Parameter sensitivity and confidence at $t = 5$ s (left) and $t = 20$ s (right).

Independently from this analysis, the Gen-II network architecture had been modified several times during the design of the controller and the number of weights in the network (for roll axis) has been reduced from 60 (as shown) to 6. Our parameter sensitivity tool provided independent statistical evidence to support this drastic change in the network architecture.

3.5. *Evaluating the Network's Output Quality*

The statistical evaluation of a neural network's prediction performance at the output layer of a neuro-adaptive system is equally important as the evaluation of its learning performance. It verifies the adaptation performance at the output layer and prevents the unreliable output from entering the next component, which usually is the controller/actuator of the system. It measures the "trustworthiness" or "confidence" of the output and alerts operators when an output is considered potentially hazardous in terms of "reliability and trustworthiness".

3.5.1. *Validity Index for DCS Networks*

Following the definition of Validity Index (VI) in RBF networks by Leonard *et al.*,[9] we define the validity index in DCS networks as an estimated confidence measure of a DCS output, given the current input. The VI can be used to measure the accuracy of the DCS network fitting and thus provide inferences for future validation activities. Based on the primary rules of DCS learning and properties of the network structure, we employ confidence intervals and variances to calculate the validity index in the DCS. The computation of a validity index for a given input consists of two steps: (1) compute the local error associated with each neuron, and (2) estimate the standard error of the DCS output for the given input using information from step (1). Details can be found in Refs. 32, 38.

In the case of our application of interest, the IFCS Gen-I, a domain specific threshold can be pre-defined to help verify that the accuracy indicated by the validity index is acceptable in the system context. This system performance validation step is enabled by the existence of the validity index. We have modified the DCS training algorithm to calculate the validity index. Because all needed information is present at the final step of each training cycle, we can simply calculate $s_i'^2$ for each neuron after the learning stops. When the DCS is in recall mode, the validity index is computed based on the local errors and then associated with every DCS output. We have simulated the online learning of the DCS network under a failure mode condition. Running at 20 Hz, the DCS network updates its learning data buffer (of size 200) at every second and learns on the up-to-date data set of size 200. We first start the DCS network under nominal flight conditions with 200 data points. After that, every second, we set the DCS network in recall mode and calculate the derivative corrections for the freshly generated 20 data points, as well as their validity index. Then we set the DCS network back to the learning mode and update the data buffer to contain the new data points.

Figure 10 shows the experimental results of our simulation on the failure mode condition. The left plot shows the final form of the DCS network structure at the end of the simulation. The 200 data points in the data buffer at the end of the simulation are shown as crosses in the 3-D space. The network structure is represented by circles (as neurons) connected by lines as a topological mapping to the learning data. The right plot presents the validity index, shown as error bars. The x-axis here represents the time frames. The failure occurs at $t = 5.0$ s. We compute the validity index for the data points that are generated five seconds before and five seconds after the failure occurs.

A trend revealed by the validity index in our simulations is the increasingly larger error bars after the failure occurs. At $t = 6.0$ s, the network has learned these 20 failure data points that have been generated between 5.0 s and 6.0 s. The network performance became less stable. After that, the error bars start shrinking while the DCS network adapts to the new domain and accommodates the failure. After the failure occurs, the change (increase/decrease) of the validity index varies depending on the characteristics of the failure as well as the accommodation performance of the DCS network. Nevertheless, the validity index explicitly indicates how well and how fast the DCS network accommodates the failure.

3.5.2. *Bayesian Confidence Tool for Sigma-Pi Networks*

For the Gen-II architecture, the *Confidence Tool* (CT)[39] produces a quality measure of the neural network output. Our performance measure is the probability density $p(o|x, D)$ of the network output o given inputs x, when the network has been trained with training data D. Assuming a Gaussian distribution, we use the variance σ^2 as our performance metric. A small σ^2 (a narrow bell-shaped curve) means that, with a high probability, the actual value is close to the returned value. This indicates a good performance of

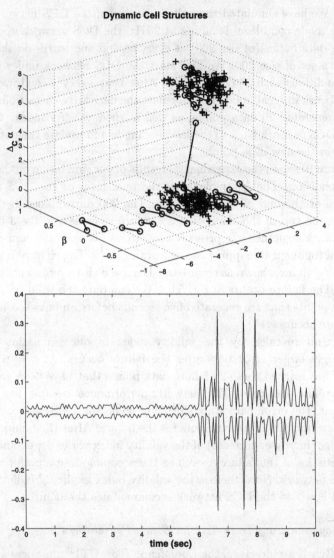

Dynamic Cell Structures

Fig. 10.   Online operation of DCS VI on failure mode simulation data. Top: The final form of DCS network structures. Bottom: VI shown as error bars for DCS output.

the network. A large σ^2 corresponds to a shallow and wide curve. Here, a large deviation is probable, indicating poor performance.

We calculate the variance σ^2 using a Bayesian approach, following the derivation in Ref. 8. The desired probability $p(\mathbf{o}|x, D)$ can be obtained by marginalizing over all possible weights w of the network:

$$p(\mathbf{o}|x, D) = \int p(\mathbf{o}|x, w)p(w|D)dw$$

The first term in the integral concerns the recall phase of the trained network, when it is subjected to input x. The second term $p(w|D)$ describes, how the weights w of the network are influenced by training it using data D. This term can be calculated as a posterior using Bayes' rule

$$p(w|D) = \frac{p(D|w)p(w)}{p(D)}$$

considering the distribution of the weights before and after the training data D has been seen by the network. Here, we use a simple Gaussian prior for $p(w)$. The probability of the data $p(D) = \int p(D|w)p(w)dw$ is a normalization factor. For the further calculations, we assume

$$p(w|D) \propto \exp\{-(\beta E_D + \alpha E_W)\}$$

where E_D is the (quadratic) training error, $E_W = \sum_i w_i^2$, and α and β are hyper-parameters. A quadratic approximation of the exponent around the most probable weights finally yields (for details see Ref. 8)

$$\sigma_t^2 = \frac{1}{\beta} + \nabla_w^T \mathbf{A}^{-1} \nabla_w$$

where ∇_w is the gradient of the network output with respect to the weights at the current input x, and $\mathbf{A} = \beta \mathbf{H}_D + \alpha \mathbf{I}$ is the regularized Hessian with respect to the network weights w. In order to keep the computationally effort during monitoring low, we chose a coarse approximation for the hyper-parameters, namely $\alpha = W/2E_W$ and $\beta = N/2E_D$ for $N \gg W$, where W is the number of weights, and N the number of training data in D.

This closed-form solution now enables the efficient calculation of our desired performance measure. The computational burden, which is mainly due to calculating the matrix inverse, could be reduced by using a moving window approach. Our confidence tool has been implemented for Sigma-Pi and multi-layer perceptron (MLP) networks in Matlab (for a Simulink environment) and in C. For details see Refs. 39, 40.

Figure 11 shows results of a (Simulink) simulation experiment for the roll axis. In the lower left panel, the pilot commands are shown over time in seconds. The pilot issues three doublet commands, which are fast stick movements from neutral into positive, then negative and

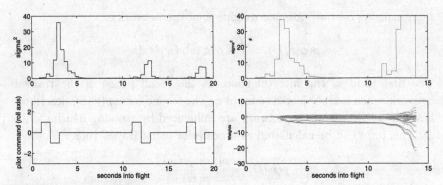

Fig. 11.   Confidence value σ^2 over time for successful and unsuccessful adaptation (*top panels*), pilot commands (*bottom left*), and diverging network weights (*bottom right*).

back to neutral position. Shortly after the first doublet ($t = 1.5\,$s), one control surface of the aircraft, the stabilizer, gets stuck at a fixed angle ("the failure"). Because the system dynamics and the model behavior do not match anymore, the neural network has to learn to produce an augmentation control signal to compensate for this deviation. This means that the network weights are updated according to the given weight update rule. Initially, the network confidence is very high, as shown in the top left panel of Fig. 11. However, as soon as the damage occurs and consequently weights are updated, the σ^2 of the network output increases substantially, indicating a large momentary uncertainty in the network output. Due to the online training of the network, this uncertainty decreases very quickly. A second and third pilot command (identical to the first one) is executed at $t = 11\,$s and $t = 17\,$s, respectively. During these commands, the network's confidence is still reduced, but much less than before. This is a clear indication that the network has successfully adapted to handle this failure situation.

The neural network's behavior in Fig. 11 (right) is in stark contrast to this successful scenario. Although the network is able to accommodate the first doublet, where the failure occurred, it is not able to continuously handle this situation: as soon as the second doublet is commanded, the network still tries to learn the new situation, but fails. The rapid (and unbounded) increase of the weight values (lower right panel) is a clear indication of the network diverging and ultimately causing unstability of the aircraft ($t = 14\,$s). The confidence value σ^2 (top right panel) clearly indicates this situation: the value of σ^2 is growing extremly fast and to very high values. Since the bad quality of the neural network can be recognized roughly 1–2 seconds prior to network divergence, the dynamic confidence tool could be used as a warning device for the pilot.

4. Conclusions

While neuro-adaptive systems hold a great promise in autonomous systems and control applications, they are problematic for verification and validation. The reason is that a neuro-adaptive system evolves over time and thus the validation techniques applied before its online adaptation are no longer applicable to the changed configuration. Furthermore, the validation of the neural network models is particularly challenging due to their complexity and nonlinearity. Reliability of learning, performance of convergence and prediction is hard to guarantee. The analysis of traditional controllers, which have been augmented by adaptive components require technically deep nonlinear analysis methods.

We developed a non-conventional approach for validating the performance of a neural network-based online adaptive system. The validation framework consists of:

- Independent failure detections at the input layer and performance assessment checks at the output layer that provide validation inferences for verifying the accommodation capabilities of the online adaptive component in the context of failure accommodation, and
- Runtime learning performance monitors that examine the internal properties of the neural network adaptation.

We developed validation techniques to examine: (1) the learning data on which the online adaptive component is trained, (2) the online adaptation performance, and (3) the neural network predictions after the adaptation. At the input layer, we have presented SVDD as a novelty detection tool for defining nominal performance regions for the given application domain and thus used for failure detection. By improving its computational efficiency, our fast SVDD algorithm achieves the ability to provide successfully automated separation between potential failures and normal system events in real-time operation. We have also presented tools for the estimation of the learning and prediction performance of two different types of neural networks used in an adaptive controller. For two, highly disjoint architectures, Dynamic Cell Structures (DCS), and Sigma-Pi networks, we have shown how the network prediction performance in form of statistical error bars (validity index for DCS, network confidence for Sigma-Pi) can be calculated. The online estimation is import in control applications, where the neural network is being trained during operation. The availability of this information plays an important role for verification and validation of such a system in a safety-critical application.

Our tools are primarily designed to provide dynamic statistical evaluation on the performance of the networks. This information is vital during system verification and validation as well as for in-flight monitoring. Furthermore, these tools can also be used during the early design phase of an adaptive controller, when the architecture and size of the network is determined. Our Bayesian approach allows different models (e.g., networks with different numbers of hidden units, or different network types such as MLP, Sigma-Pi, RBF, or DCS) to be compared using only the training data. More generally, the Bayesian approach provides an objective and principled framework for dealing with the issues of model complexity.

References

1. Rysdyk, R., Calise, A.: Fault tolerant flight control via adaptive neural network augmentation. *AIAA-98-4483*, 1998, pp. 1722–1728.
2. Norgaard, M., Ravn O., Poulsen, N., Hansen, L.K.: *Neural Networks for Modeling and Control of Dynamic Systems*. Springer, 2002.
3. Jorgensen, C.C.: Feedback linearized aircraft control using dynamic cell structures, *World Automation Congress (ISSCI)*, Alaska, 1991, pp. 050.1–050.6.
4. The Boeing Company, Intelligent flight control: Advanced concept program, *Technical Report*, 1999.
5. Boyd, M.A., Schumann, J., Brat, G., Giannakopoulou, D., Cukic, B., Mili, A.: Validation and verification process guide for software and neural nets. Technical Report, NASA Ames Research Center, 2001.
6. Schumann, J., Nelson, S.: Towards V&V of neural network based controllers. *Workshop on Self-Healing Systems*, 2002.
7. Mackall, D., Nelson, S., Schumann, J.: Verification and validation of neural networks of aerospace applications. Technical Report CR-211409, NASA, 2002.
8. Bishop, C.M.: *Neural networks for pattern recognition*, Oxford University Press, Oxford, UK, 1995.
9. Leonard, J.A., Kramer, M.A., Ungar, L.H.: Using radial basis functions to approximate a function and its error bounds, *IEEE Transactions on Neural Networks*, 3(4), 1992, pp. 624–627.

10. Fuller, E., Yerramalla, S., Cukic, B., Gururajan, S.: An approach to predicting non-deterministic neural network behavior. In: Proc. Intl. Joint Conference on Neural Networks (IJCNN), 2005.
11. Mili, A., Cukic, B., Liu, Y., Ben Ayed, R.: Towards the verification and validation of on-line learning adaptive systems, In *Computational Methods in Software Engineering*, Kluwer Scientific Publishing, 2003.
12. Hornik, K.M., Stinchcombe, M., White, H.: Multilayer feedforward networks are universal approximators, *Neural Networks*, 2, 1989, pp. 359–366.
13. Raz, O.: Validation of online artificial neural networks — an informal classification of related approaches. Technical report, NASA Ames Research Center, Moffet Field, CA, 2000.
14. Del Gobbo, D., Cukic, D..: Validating on line neural networks Technical Report, Lane Department of Computer Science and Electrical Engineering, West Virginia University, December 2001.
15. Institute of Software Reseach, Dynamic cell structure neural network report for the intelligent flight control system, *Technical Report*. Document ID: IFC-DCSR-D002-UNCLASS-010401, January, 2001.
16. Darrah, M., Taylor, B., Skias, S.: Rule extraction from Dynamic Cell Structure neural networks used in a safety critical application, *Proc. of the 17th International Conference of the Florida Artificial Intelligence Research Society*, September, 2004.
17. Yerramalla, S., Fuller, E., Cukic, B.: Lyapunov analysis of neural network stability in an adaptive flight control system, *6th Symposium on Self-Stabilizing Systems (SSS-03)*, San Francisco, CA, June 2003.
18. Yerramalla, S., Cukic, B., Fuller, E.: Lyapunov stability analysis of quantization error for DCS neural networks, *Int'l Joint Conference on Neural Networks (IJCNN'03)*, Oregon, July, 2003.
19. Tibshirani, R.: Bias, variance and prediction error for classification rule. Technical Report, Statistics Department, University of Toronto, 1996.
20. Fu, L.: Neural Networks in Computer Intelligence, *McGraw Hill*, 1994.
21. Peterson, G. E.: A foundation for neural network verification and validation, *SPIE Science of Artificial Neural Networks II*, 1966:196–207, 1993.
22. Hunt, K.J., Sbabaro, D., Zbikowski, R., Gawthrop, P.J.: Neural networks for control systems — A survey, *Automatica*, 28(6), 1996, 1707–1712.
23. Lawrence, S., Tsoi, A.C., Back, A.D.: Function approximation with neural networks and local methods: Bias, variance and smoothness, *Australian Conference on Neural Networks*, Peter Bartlett and Anthony Burkitt and Robert Williamson, 1996, 16–21.
24. Grossberg, S.: Adaptive pattern classification and universal recoding: I. Parallel development and coding of neural feature detectors. *Biological Cybernetics*, 23:121–134, 1976. Reprinted in Anderson and Rosenfeld. (1988).
25. Grossberg, S.: Competitive learning: From interactive activation to adaptive resonance, *Cognitive Science*, 11(1), January–March 1987, 23–63.
26. Schumann, J., Gupta, P.: Monitoring the performance of a neuro-adaptive controller. In: Fischer, R., Preuss, R., von Toussaint, U.: Proc. 24th

International Workshop on Bayesian Inference and Maximum Entropy Methods in Sciences and Engineering (MAXENT), AIP 2004, 289–296

27. Kohonen, T.: Self-Organizing Maps, Springer-Verlag, New York, 1997.

28. Martinez, T., Schulten, K.: Topology representing networks, Neural Networks, 7(3), 1994, pp. 507–522.

29. Bruske, J., Sommer, G.: Dynamic cell structures, In: Proc. Neural Information Processing Systems, Vol. 7, 1995, pp. 497–504

30. Ahrns, I., Bruske, J., Sommer, G.: On-line learning with dynamic cell structures. In: Fogelman-Soulié, F., Gallinari, P. (eds.): Proc. Int. Conf. Artificial Neural Networks, EC2, Nanterre, France, Vol. 2. 1995, 141–146.

31. Fritzke, B.: Growing cell structures — A self-organizing network for unsupervised and supervised learning, Neural Networks, 7(9), 1993, pp. 1441–1460.

32. Liu, Y.: Validating a neural network-based online adaptive system. Ph.D. thesis, West Virginia University, Morgantown, 2005.

33. Rumelhart, McClelland, and the PDP Research Group, Parallel distributed processing. MIT Press, 1986.

34. Tax, D.M.J., Duin, R.P.W.: Support vector domain description, Pattern Recognition Letters, 20(11–13), 1999, 1191–1199.

35. Tax, D.M.J., Duin, R.P.W.: Data domain description using support vectors, Proc. European Symposium on Artificial Neural Networks (Bruges, April 21–23, 1999), D-Facto, Brussels, 1999, pp. 251–257.

36. Tax, D.M.J.: One-class classification, Dissertation, ISBN: 90-75691-05-x., 2001.

37. Bennett, K.P., Campbell, C.: Support vector machines: Hype or hallelujah? SIGKDD Explorations, Vol. 2.2., 2000, pp. 1–13.

38. Liu, Y., Cukic, B., Jiang, M., Xu, Z.: Predicting with confidence — An improved dynamic cell structure. In: Wang., L, Chen, K., Ong., Y.S. (eds.): Lecture Notes in Computer Science: Advances in Neural Computation, Springer-Verlag, Berlin Heidelburg, Vol. 1, 2005, 750–759.

39. Gupta, P., Schumann, J.: A tool for verification and validation of neural network based adaptive controllers for high assurance systems. In: Proc. High Assurance Software Engineering, IEEE Press (2004).

40. Schumann, J., Gupta, P., Jacklin, S.: Toward verification and validation of adaptive aircraft controllers. In: Proc. IEEE Aerospace Conference, IEEE Press (2005).

Chapter 3

ADAPTIVE RANDOM TESTING

DAVE TOWEY

BNU–HKBU United International College
28, Jinfeng Road, Tangjiawan, Zhuhai
Guangdong Province 519085, China
davetowey@uic.edu.hk
http://www.uic.edu.hk/~davetowey

Computing is everywhere, and software is too, but what about the quality of this software? As software becomes increasingly pervasive in our society, what can we do to ensure its quality? One Software Quality Assurance mechanism, Software Testing, is often omitted from the development and implementation of computer systems, sometimes because of its perceived inconvenience and difficulty. A relatively simple way of testing software is to apply test cases (combinations of input representing a single use of the software) randomly, a method known as Random Testing. Among the advantages of Random Testing are its ease of use, the minimal overheads in test case generation, and the statistical support available. Some research has indicated that more widespread distributions of test cases throughout the input domain may be more effective at finding problems in the software. Adaptive Random Testing methods are Software Testing methods which are based on Random Testing, but which use additional mechanisms to ensure more even and widespread distributions of test cases over an input domain. This chapter gives an introduction to some of the major Adaptive Random Testing implementations.

1. Introduction

Although it has been a long time since Naur first introduced the phrase "Software Engineering",[47] at a NATO conference in 1968 to address the then perceived crisis in software production,[51] even now Software Engineering *(SE)* is often considered by many to be an emerging discipline.[1,50] Indeed, many would argue that little has really changed, and that there is still a software crisis[37] — a recent article estimates that US$60 billion to $75 billion is wasted each year due to software problems.[2] As computing and software become more ubiquitous, it seems that software problems are also likely to increase. An obvious area of concern therefore, will be the quality of the software running the computing!

Methods for improving or ensuring Software Quality, referred to as Software Quality Assurance (*SQA*) techniques, are often classified into dynamic and static methods.[1,35] The distinction refers to whether the code is being executed (dynamic methods), or not (static methods). Software Testing is a dynamic method of *SQA*.

Although there was a period when almost all literature on Software Testing included a lament on its position as an under-used, often-omitted, and poorly understood part of the software process,[57] this situation is changing. Newer approaches to software development now even advocate writing tests before the code.[4-6,8] It has also been said that of the different disciplines in software development, Software Testing is perhaps the most mature.[3]

Possible reasons for the lack of Software Testing, in some situations, include the difficulty associated with the preparation, or the amount of time necessary to implement the testing. If the testing process were to be made simpler, or even automated,[27,39] it is probable that it may be more frequently used.

Fundamental to testing software is the test case,[45] a collection of inputs to the Software Under Test (*SUT*) which represent a single execution of the software. Since exhaustive testing of all possible inputs to the *SUT* is usually prohibitively difficult or expensive, it is essential for testers to make the best use of their limited testing resources. To do this, various testing (test case selection) methods have evolved. Testing methodologies have often been categorized broadly into two types: those that make use of information concerning the program's structure in the preparation of test cases (called White Box, or Structural Testing); and those that do not (called Black Box, or Functional Testing).[45,49] There has also been a growing trend to call certain approaches Grey, or Broken-Box techniques, where some (relatively small) amount of information about the program's structure is used.[55]

A simple Black Box method of testing software is to draw test cases at random from the input domain and apply them to the Software Under Test, a method referred to as Random Testing (*RT*). It seems that few topics in Software Testing are as controversial as the issue of whether or not *RT* should be used.[7] Myers called it "the poorest [test case design] methodology of all",[46] but later studies showed that it may be worthwhile, and even more cost effective for many programs.[25,26] A method with which *RT* is often compared is Partition Testing, where the input domain is partitioned into subdomains prior to selecting test cases from each, randomly or according to some guidelines.[10,30] As Chan *et al.* state, intuitively speaking, Partition Testing should be more effective in revealing program errors than Random Testing,[10] yet many studies suggest that this may not be so.[22,25,26,29,31,56]

There are many reasons why Random Testing is a popular choice: in addition to its simplicity, and the efficiency of test case generation,[38] many reliability estimates and statistical analyses are also easily performed.[26,32–34,54] In spite of some controversy over its effectiveness, many real-life applications do make use of Random Testing.[28,42,43,48,52,58] It has also been suggested that Random Testing may be a good option at the end of other forms of testing,[53] or even instead of them.[41]

Although in real testing situations, duplication of test cases is unnecessary, or even undesirable, for testing strategies where the cost of checking for duplication is higher than the cost of test case execution (as is usually the case for *RT*), it is quite common to allow duplication. In addition to computational simplicity, this replacement of test cases gives rise to simpler mathematical models, and thus facilitates the analysis of testing strategies.[36]

A failure pattern for a program is defined as the collection of all points within the input domain which, if executed, would reveal a fault or cause the Software Under Test to fail. Chan *et al.*[9] reported on how different failure patterns could influence the performance of some testing strategies. They identified three major categories of failure patterns: *point*, characterized by individual or small groups of failure-causing input regions; *strip*, characterized by a narrow strip of failure-causing inputs; and *block*, characterized by the failure-causing inputs being concentrated in either one or a few regions. Examples of each of these are given in the schematic diagrams in Figs. 1a to 1c, where the outer boundaries represent the borders

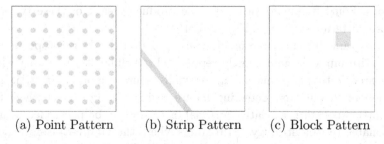

(a) Point Pattern   (b) Strip Pattern   (c) Block Pattern

Fig. 1.   Examples of Failure Pattern Types.

```
int X, Y, Z;
cin >> X >> Y;
if ( ((X MOD 4) == 0)
    AND
    ((Y MOD 4) == 0) )
{
    Z = function_1(X, Y);
    /* CORRECT CODE IS
        Z = function_3(X, Y);
    */
} else {
    Z = function_2(X, Y);
}
cout << Z;
```

Fig. 2.   Code fragment producing the Point Failure Pattern Type.

```
int X, Y, Z;
cin >> X >> Y;
if ( X + Y > 12)
/* CORRECT CODE IS
    if ( X + Y > 10)
*/
{
    Z = function_1(X, Y);
} else {
    Z = function_2(X, Y);
}
cout << Z;
```

Fig. 3.   Code fragment producing the Strip Failure Pattern Type.

of the input domain, and the shaded areas represent the failure-causing regions. Some fragments of C++ code producing each of these types of failure pattern are given in Figs. 2 to 4.

Programs in which the failure-causing region is a large proportion of the entire input domain usually represent little difficulty in testing when compared with programs having a small failure region. In fact, it has been pointed out that detecting failure regions for programs where the proportion is high is a relatively trivial task, and can be performed by any reasonable testing strategy.[17] Programs where the failure-causing regions are comparatively small represent a greater challenge to Software Testing.

```
int X, Y, Z;
cin >> X >> Y;
if ( (X > 10) AND (X < 20) )
   AND
   ( (Y > 30) AND (Y < 40) )
{
    Z = function_1(X, Y);
    /* CORRECT CODE IS
        Z = function_3(X, Y);
    */
} else {
    Z = function_2(X, Y);
}
cout << Z;
```

Fig. 4.   Code fragment producing the Block Failure Pattern Type.

2. Adaptive Random Testing

Chan *et al.*[9] observed that the performance of some testing strategies may be influenced by the pattern of the failure-causing inputs in the input domain (the failure pattern). This prompted investigation into improving performance of Random Testing by incorporating information about failure patterns. Methods based on Random Testing, but involving additional strategies to take advantage of failure pattern insights, have been named Adaptive Random Testing (*ART*) methods.[19,20,23] These methods have been identified as a very promising direction for automatic testing.[24]

An insight from the *ART* research is that a more widespread or even distribution of the test cases over the input domain may be more favorable for failure-finding. The intuition underlying why this should be so may be explained by means of a simple example: Consider a two-dimensional input domain with a circular failure region, the center of which is O, and the radius of which is r (Fig. 5) Suppose a test case t is randomly generated to test the program, but does not reveal a failure (that is, t falls outside the failure region). Although both the location of O and the value of r are unknown, O is clearly at a distance of at least of r from t. Obviously, any test case drawn from the circular failure region is sufficient to show that the program is faulty, but for illustration of the intuition, assume that the testing objective is to select O as a test case. In view of the absence of the knowledge of r, intuitively speaking, it is better to choose a test case far

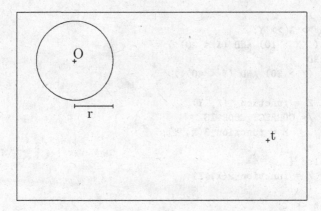

Fig. 5.   Illustration of widespread distribution intuition.

away from t rather one close to t. Since the input domain is bounded, "far away" will effectively mean "widespread" or "evenly distributed".

There are many ways by which test case selection can be encouraged to be more widespread, and hence there are many possible implementations of ART. In the next sections, some of the major ART methods will be introduced.

2.1. *Distance-Based Adaptive Random Testing*

Distance-based ART ($DART$) is an ART strategy based on maximizing minimum distances among test cases.[20,21] $DART$ makes use of two sets of test cases: the executed set, a set of test cases which have been executed but without causing failure; and the candidate set, a set of cases selected randomly from the input domain. The executed set is initially empty and the first test case is randomly generated from the input domain. Each time a test case is required for execution, the element in the candidate set that is *farthest away* from all the executed test cases is selected. When examining the elements to determine which is *farthest away*, the Euclidean distance is calculated, then test cases are selected such that those with the maximum minimum distance from the previously executed test cases are selected.

One version of $DART$ is the Fixed Size Candidate Set (FSCS) version, in which the candidate set is maintained with a constant number of elements. Each time an element is selected and executed, the element is added to the executed set, and the candidate set is completely reconstructed.

Experiments have revealed that the failure finding efficiency of FSCS improves as the size of the candidate set increases.[20,21]

2.2. *Restriction-Based Adaptive Random Testing*

Another version of *ART*, based on the restriction of eligible regions for test case selection, is Restricted Random Testing,[11,19] also known as Restriction-based *ART* (*RART*). By excluding regions surrounding previously executed test cases, and restricting subsequent cases to be drawn from other areas of the input domain, *RRT* ensures an even distribution, and guarantees a minimum distance amongst all cases.

When testing according to the *RRT* method, given a test case that has not revealed failure, rather than simply select another test case randomly, the area of the input domain from which subsequent test cases may be drawn is restricted. In two dimensions, a circular exclusion zone around each non-failure-causing input is created, and subsequent test cases are restricted to coming from outside of these regions. By employing a circular zone, a minimum distance (the radius of the exclusion zone) between all test cases is ensured.

All exclusion zones are of equal size, and this size decreases with successive test case executions. The size of each zone is related to both the size of the entire input domain, and the number of previously executed test cases. For example, in two dimensions, with a target exclusion region area of A, if there are n points around which we wish to generate exclusion zones, then each exclusion zone area will be A/n, and each exclusion zone radius will be $\sqrt{A/(n\pi)}$.

Figure 6 shows a graphical representation of the generation of the first few test cases using the *RRT* method. As shown, after each test case is generated (and presumably applied to the program without revealing failure), exclusion zones are created around all non-failure-causing test cases, and the next test case is selected from outside these excluded regions.

The size of each exclusion zone is related to both the size of the entire input domain, and the number of previously executed test cases. The final (and most important) determinant of exclusion zone size is the Exclusion Ratio (R).[14] This figure, represented as a fraction, is applied to the total area of the input domain to obtain the desired total exclusion area. For example, in a two-dimensional input domain of total area D, for an Exclusion Ratio R, the target exclusion region area A is RD; with n

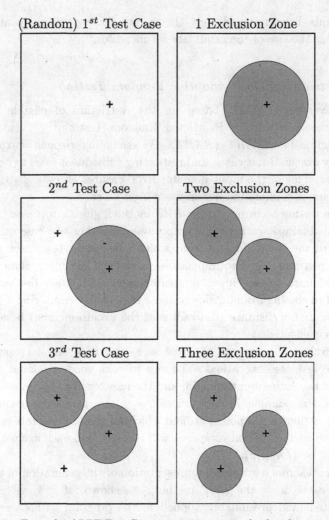

Fig. 6. Example of RRT Test Case generation process for first few test cases.

exclusion zones, each exclusion zone area will be RD/n, and each exclusion zone radius will be $\sqrt{RD/(n\pi)}$.

Figure 6 showed situations with one, two, and three exclusion regions. In this example, the total exclusion region area remained constant, that is, the area for one exclusion region was the same as the sum of the areas of the two exclusion regions, which was also the same as the sum of the areas of the three exclusion regions. This is as expected; the Actual Exclusion Ratio is equal to the Target Exclusion Ratio.

Further research into the *RRT* method and the Exclusion Ratio (*R*) revealed that, many times during the execution of the *RRT* algorithm, the Actual Exclusion Ratio is less than the Target Ratio.[14] Furthermore, it was discovered that the *RRT* method performed best when the value used for the Target Exclusion Ratio (*R*) was maximized, a value referred to as the Maximum Target Exclusion Ratio (*Max R*).

By varying the size and shape of the input domains in simulations using *RRT*, it was discovered that these parameters affect the value of the Maximum Target Exclusion Ratio (*Max R*). This led to the development of a normalizing feature to be added to the Ordinary *RRT* (*ORRT*) resulting in the Normalized version of Restricted Random Testing (*NRRT*).[17,19] *NRRT* normalized the input domain and made it possible to approximate the *Max R*, depending on the dimensionality of the Software Under Test (*SUT*).

Both the *ORRT* and *NRRT* versions of Restricted Random Testing performed very well in empirical studies, displaying improvements in performance as the values for the Target Exclusion Ratio (*R*) were increased, and best performance when *R* was maximized (*Max R*).[11,17,19]

Because the distance calculations associated with circular exclusion zones can become computationally quite expensive, and because of the complexity of the relationship between the Target and Actual Exclusion ratios (due, in part, to the circular shape of the exclusion region), an alternative implementation of *RRT* with square exclusion shapes was investigated.[12,19]

An advantage of the square exclusion shape is the cheaper inequality operations used to verify if a test case lies within an exclusion region (compared with the distance calculations necessary for a circular exclusion shape).

Results of experiments using the square exclusion shape revealed similar trends to those for the circular exclusion versions of *RRT*, but poorer overall results.[12,19]

2.3. *Overheads*

The *RRT* methods incur potentially significant overheads in the generation of the $(m + 1)$th test case. At this instant, when generating the $(m + 1)$th test case, there are already m exclusion regions around m executed test cases, and the $(m + 1)$th test case is restricted to coming from outside these regions. A simple implementation of the exclusion region is to ensure that the candidate test case is a greater distance from each executed test case than the radius of the exclusion region. For two points, P and Q $((p_1, p_2, \ldots, p_N)$ and $(q_1, q_2, \ldots, q_N))$, the Euclidean distance between the points can be calculated from the following expression:

$$\text{Distance}(P, Q) = \sqrt{\sum_{i=1}^{N} (p_i - q_i)^2} \tag{1}$$

Ignoring possible optimizations, in a best case scenario, where the first candidate test case is outside all exclusion regions, there are m distance calculations required to confirm that the $(m + 1)$th test case is acceptable. In practice, it is possible that several attempts at generating an acceptable test case will be required. For each unacceptable candidate, there will have been x number of comparisons (and hence x distance calculations) prior to that comparison revealing the test case to be within an exclusion region. The value of x will be between 1 and m, the worst case being that the candidate is found to be within the final exclusion region checked.

2.4. *Filtering*

Motivated by the lower computation overheads of the square exclusions (Section 2.2), a hybrid approach, called *Filtering*, was developed.[12] In this approach, given a candidate test case which is being examined to verify that it does not lie within any exclusion region, a bounding square/cube/hypercube is established around this test case, and used to filter the previously executed test cases, calculating the distance only for those test cases inside the Bounding Region. The Bounding Region corresponds to a square/cube/hypercube restriction zone, requiring only the cheaper inequality operation. The number of test cases that lie inside the Bounding Region is significantly less than the total number in the entire input domain. With normal distribution of test cases, the number expected to fall inside the bounding region is proportional to the relative size of the bounding region. In the following equations, BRP is the expected Bounding Region Population, i.e., the number of test cases expected to be in the Bounding Region; and m is the number of executed test cases, which is also the number of exclusion regions.

$$BRP = m \times \frac{\text{Size}_{\text{Bounding Region}}}{\text{Size}_{\text{Input Domain}}} \tag{2}$$

The magnitude of the Bounding Region side is twice that of the exclusion region radius.

$$\text{Size}_{\text{Bounding Region}} = [2r]^N \qquad (3)$$

The value of the exclusion radius (r) depends on the size of each exclusion region, which in turn depends on the size of the entire input domain, the Exclusion Ratio (R), and the total number of exclusion regions (m).

$$\text{Size}_{\text{Exclusion Region}} = \frac{R \times \text{Size}_{\text{Input Domain}}}{m} \qquad (4)$$

The formula to calculate the radius changes according to the dimensions of the input domain. Table 1 summarizes the expected number of test cases to fall inside a bounding region for 2, 3 and 4 dimensions.

Experiments have verified the filtering method's speed compared with the ordinary, circular exclusion implementation of *RRT*, while maintaining identical failure-finding results. As Table 1 shows, the (maximum) number of test cases on which the more expensive circular exclusion method will be applied is a small constant, determined by the size of the exclusion regions: for example, with Target Exclusion Ratio (R) of 150%, in two dimensions, it is expected to be applied to less than two, all other test cases being filtered.

Obviously, when larger proportions of the input domain are excluded, it is more difficult to randomly generate a test case lying in a non-excluded area. In this case, the number of attempts to generate an acceptable test

Table 1. Expected number of test cases falling in Bounding Region, for 2, 3, and 4 dimensions. A is the total input domain area/volume; m is the number of executed test cases; R is the Exclusion Ratio; and r is the radius of the exclusion regions.

N	Area/Volume/ Hyper Volume (for circle, sphere, etc)	Expected number of test cases inside Bounding Region
2	$\pi r^2 = \frac{A \times R}{m}$	$\frac{4R}{\pi}$ E.g., $R = 150\%$ means less than 2 test cases
3	$\frac{4}{3}\pi r^3 = \frac{A \times R}{m}$	$\frac{6R}{\pi}$ E.g., $R = 270\%$ means less than 6 test cases
4	$\frac{1}{2}\pi^2 r^4 = \frac{A \times R}{m}$	$\frac{32R}{\pi^2}$ E.g., $R = 430\%$ means less than 14 test cases

case increases. Approximately, if 99% of the input domain is excluded, leaving only 1% from which the test case may be drawn, it should take an average of 100 (1/1%) attempts to generate an acceptable test case.

Filtering enables a significant reduction in the overheads associated with the generation of an acceptable test case while maintaining the failure-finding efficiency of the basic methods.[12]

2.5. *Forgetting*

Human learning is often characterized by inaccurate retention or recall, termed forgetting, which as Markovitech and Scott pointed out, "is usually regarded as an unfortunate failure of the memory system".[40] Markovitech and Scott explored the potential benefits of such failure in the context of Machine Learning,[44] finding that in addition to the obvious reductions in overheads, even random deletion of knowledge yielded improvements in system performance.

In the context of Adaptive Random Testing (*ART*), *Forgetting* refers to modifications to the basic methods which result in the original algorithm using only some of the available test cases, "forgetting" the rest. For *RRT*, *Forgetting*[15] is motivated mainly by the desire to reduce overheads. A feature of the *RRT* method is that as the number of executed test cases increases, the size of each individual exclusion zone decreases. This allows the selection of new test cases to be increasingly close to previously executed test cases, as we want, but it also increases the computational burden; the exclusion radius decreases in size such that at its extreme, when the radius is of negligible length, we are effectively performing Random Testing, but with considerably higher overheads. If *Forgetting* is applied in such a way that, for example, only a maximum of k test cases were used in the *RRT* algorithm, then the minimum size of exclusion zones would be known in advance, and some assurance of the cost-benefit trade-off would be available.

Three implementations of *Forgetting* were investigated: *Random Forgetting*; *Consecutive Retention*; and *Restarting*. *Random Forgetting* refers to an implementation where, at the time of generating the $(m + 1)$th test case, $m - k$ executed test cases are randomly selected and deleted, and the *RRT* algorithm is applied only to the remaining k; the *Consecutive Retention* implementation deletes the first $m - k$ test cases, retaining the last, consecutive, k test cases; the *Restarting* implementation involves a complete reset of the algorithm — after k test cases, and exclusion regions implemented as usual, *Restarting* entirely *forgets* everything that has happened, and restarts the *RRT* method.

Results from experiments showed that the *Forgetting* methods could perform similarly to the basic *RRT* methods, including yielding better results as the Target Exclusion Ratio (R) increased.[15]

2.6. *Mirror* ART

Chen *et al.*[23] introduced a testing methodology called *Mirroring* which can be combined with other testing strategies to reduce computational overheads. In addition to the reduction of computational costs, *Mirroring* offers several other interesting properties.

To apply *Mirroring*, the input domain of the Software Under Test (SUT) is partitioned into p disjoint subdomains. The test case generation algorithm is performed in only one of the subdomains, designated the source subdomain. After generation, if the test case does not reveal a failure in the source subdomain, it is mapped (using a mirror function) to the other subdomains, known as mirror subdomains, and its image applied to the SUT. If no failure has been revealed after the $p - 1$ mirror images of the source test case have been executed, the algorithm is again applied in the source subdomain to generate the next test case.

The partitioning scheme used when applying *Mirroring* is referred to as *Mirror Partitioning*, and is usually referenced according to the dimensions of the input domain, and the number of partitions on each dimension. For example, in three dimensions (X, Y, Z), where there are 2, 3, and 4 equal partitions on the respective dimensions, the *Mirror Partitioning* schema is X2Y3Z4. Figure 7 gives some further examples, in two dimensions, of *Mirror Partitioning*.

One obvious potential application of *Mirroring* is in those cases where the input domain is not regular. As explained in Section 2.2, the shape

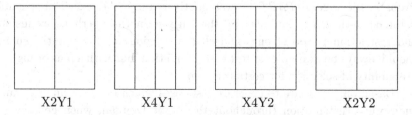

X2Y1   X4Y1   X4Y2   X2Y2

Fig. 7.   Examples of Mirror Partitioning.

of the input domain affects the performance of the $ORRT$ algorithm; investigations have revealed that applying the optimal parameter values in less regularly shaped input domains is problematic.[14,19] With $Mirroring$, it should be possible to partition such input domains so as to create a relatively homogeneous source subdomain, thus allowing the $ORRT$ method to be applied with good confidence that the optimal value of R ($Max\ R$) is used.

Because Adaptive Random Testing incorporates additional information into the test case selection, it incurs additional overheads, particularly in comparison with ordinary Random Testing (RT). $Mirroring$, by imposing the m partitions, alleviates a considerable amount of the computational overheads of whatever test case generation strategy it is applied to. An approximation of the cost reductions with $Mirroring$ suggests that overheads in test case generation can be reduced to a fraction of those of the original method: for example, with m partitions, $Mirroring$ would require only $1/m$ as many test cases to be generated by the original method to produce the same total number of test cases.

A second attractive feature of $Mirroring$ is, in addition to the reduction of computational overheads, the possibility of homogenizing, to some degree, an input domain. As previously explained, for the RRT methods, the control parameter, the Exclusion Ratio (R), yields best failure detection rates when the $Max\ R$ value is used.[14] Because the $Max\ R$ is more easily estimated in regular input domains, the potential homogenizing effect on an input domain by Mirror Partitioning is very attractive for the $ORRT$ method.

In analyses of $Mirroring$ applied to $DART$ and RRT, it was discovered that similar failure-finding efficiency could be obtained, but with considerable reduction in the computations involved.[16,23]

2.7. Probabilistic ART

Probabilistic ART ($PART$)[13] was motivated by the excellent results obtained with RRT, but also by the suggestion that perhaps exclusion and restriction is too strong, and that all regions of the input domain should always be available for test case generation, but with a bias or higher probability of selection for certain regions.

Based on the desire to generate test cases which result in an even and widespread distribution throughout the Input Domain, when generating the ith test case, we would like for it to be as far as possible away from previous test cases. According to the basic *RRT* method, this is achieved by implementing restriction zones around the previously executed test cases, and ensuring that the *i*th case is generated from outside of all these regions. The Probabilistic methods, instead of using strict exclusion, attempt to generate the *i*th test case by biasing the probability of selection such that regions which would be inside the restriction zone are proportionately less likely to be selected than regions further away.

Two implementations of *PART* were investigated: *Probabilistic Translation* and *Probabilistic Generation*. The *Probabilistic Translation* method is based on the idea of generating a random vector and selecting a new test case by translating a randomly selected previous one. The *Probabilistic Generation* method creates a map of probabilities of selection throughout the Input Domain, with selection probability close to previously executed test cases being increasingly less than areas further away.

Although further study is required for the *PART* methods, initial results were positive, indicating good improvement over Random Testing.[13]

2.8. *Fuzzy ART*

From the insights gained in earlier studies,[11,12,17,20,21] it was possible to distinguish what might make a *good* test case from a less *good* one. In particular, given information about the input domain of the Software Under Test (*SUT*), and the current set of executed test cases, it is possible to evaluate which potential test cases will contribute more to the goal of having a well-distributed test case selection pattern than others. Fuzzy set theory offers a succinct methodology within which this can be framed. This implementation of *ART*, which uses Fuzzy Set Theory to ensure the even spread of test case, is called Fuzzy Adaptive Random Testing (*FART*).[18]

In order to minimize the number of test cases required to find a failure in the *SUT*, it is desirable that the tester be able to select *good* test candidates. The definition of *good* here is restricted by the fact that it is Black Box Testing being conducted, and hence nothing is known about the internal workings of the *SUT*. What is available however, is the information that, for certain types of failure patterns (non-point type), a widespread, even distribution of test cases is most efficient at finding failure. *Good* can therefore be quantifiable in terms of how much a test case will contribute to the goal of a widespread and distributed test case pattern.

Fuzzy Set Theory[59] permits membership of a set to be defined in terms of a degree of belonging, and subsumes the crisp approach of discrete membership, or non-membership, found in traditional set theory. According to the theory, elements no longer need be classified as either "in" or "out", but can instead be given a grade of membership (usually between 0 and 1) of the set.

Considering all the potential test cases within an input domain as a set, effectively an infinite set, then it is possible to conceive of a subset containing elements that are considered *good*. The members of this *good* subset would change according to the pattern of executed test cases, but a function giving the degree of membership of this *good* subset can be designed. In particular, features that have been shown to encourage a widespread and even distribution of test cases over the input domain are evaluated and incorporated into the function.

Investigations into the performance of *FART* showed it to considerably outperform Random Testing, and compare well with other *ART* methods.[18]

3. Summary

Computing and software have become more and more significant in our lives, leading to concerns about the quality of both!

One approach to ensuring Software Quality is Software Testing, where software is executed with the intention of finding problems or errors. Simplifying ways of conducting Software Testing may increase the amount of testing done. A simple method of Software Testing, called Random Testing, requires only that test cases (combinations of input to the program representing a single execution) be drawn randomly from the input domain.

The collection of test cases, or points in the program's input domain, which would cause the failure of, or reveal a fault in, the software when executed, is called the program's failure pattern. Research into failure patterns revealed how certain patterns could influence the performance of some testing strategies.

Methods based on Random Testing, but involving additional strategies to take advantage of failure pattern insights, are called Adaptive Random Testing (*ART*) methods. This chapter has introduced some of the major implementations of *ART*.

Acknowledgements

Thank you to Prof. T. Y. Chen, Dr. K. P. Chan, Dr. R. Merkel, and Dr. F.-C. Kuo for all the support and lively discussions.

References

1. SWEBOK — Guide to the Software Engineering Body of Knowledge — 2004 Version. http://www.swebok.org/, 2004. [Online; accessed 12-July-2008].
2. Spectral Lines: Learning from Software Failure, *Spectrum, IEEE*, 42(9), 2005, p. 8.
3. Phillip, G. Armour.: Not-Defect: The mature discipline of testing, *Communications of the ACM*, 47(10), 2004, pp. 15–18.
4. Kent Beck. *Extreme Programming Explained: Embrace Change*. Addison-Wesley Longman Publishing Co., Inc., Boston, MA, USA, 2000.
5. Kent Beck. *Test Driven Development: By Example*. Addison-Wesley Longman Publishing Co., Inc, Boston, MA, USA, 2002.
6. Kent Beck, Mike Beedle, Arie van Bennekum, Alistair Cockburn, Ward Cunningham, Martin Fowler, James Grenning, Jim Highsmith, Andrew Hunt, Ron Jeffries, Jon Kern, Brian Marick, Robert C. Martin, Steve Mellor, Ken Schwaber, Jeff Sutherland, and Dave Thomas. Manifesto for Agile Software Development. http://agilemanifesto.org/, 2001. [Online; accessed 12-July-2008].
7. Boris Beizer: *Software Testing Techniques*, John Wiley & Sons, Inc., New York, NY, USA, 1990.
8. Barry Boehm and Richard Turner. Balancing Agility and Discipline: Evaluating and Integrating Agile and Plan-Driven Methods, 2004, pp. 718–719.
9. Chan, F.T., Chen, T.Y., Mak, I.K. Yu, Y.T.: Proportional sampling strategy: Guidelines for software testing practitioners. *Information and Software Technology*, 28(12), 1996, pp. 775–782.
10. Chan, F.T., Chen, T.Y., Tse, T.H.: On the effectiveness of test case allocation schemes in partition testing, *Information and Software Technology*, 39(10), 1997, 719–726.
11. Kwok Ping Chan, T.Y. Chen, Dave Towey. Restricted Random Testing. In Jyrki Kontio and Reidar Conradi, editors, *ECSQ '02: Proceedings of the 7th International Conference on Software Quality*, volume 2349 of *Lecture Notes in Computer Science*, pages 321–330, London, UK, 2002. Springer-Verlag.
12. Kwok Ping Chan, T.Y. Chen and Dave Towey. Adaptive Random Testing with Filtering: An Overhead Reduction Technique. In *17th International Conference on Software Engineering and Knowledge Engineering (SEKE'05)*, Taipei, Taiwan, Republic of China, 2005.
13. Kwok Ping Chan, T.Y. Chen and Dave Towey. Probabilistic Adaptive Random Testing. In *QSIC '06: Proceedings of the Sixth International Conference on Quality Software*, pages 274–280, Washington, DC, USA, 2006. IEEE Computer Society.
14. Kwok Ping Chan, T. Y. Chen and Dave Towey. Controlling Restricted Random Testing: An Examination of the Exclusion Ratio Parameter. In *SEKE*, pages 163–166. Knowledge Systems Institute Graduate School, 2007.
15. Kwok Ping Chan, T.Y. Chen and Dave Towey. Forgetting Test Cases. In *COMPSAC 2006: Proceedings of the 30th Annual International Computer*

Software and Applications Conference (COMPSAC 2006), Washington, DC, USA, to appear 2006. IEEE Computer Society.

16. Kwok Ping Chan, Tsong Yueh Chen, Fei-Ching Kuo, Dave Towey: A Revisit of Adaptive Random Testing by Restriction. In *COMPSAC '04: Proceedings of the 28th Annual International Computer Software and Applications Conference (COMPSAC'04)*, pages 78–85, Washington, DC, USA, 2004. IEEE Computer Society.

17. Kwok Ping Chan, Tsong Yueh Chen, Dave Towey: Normalized Restricted Random Testing. In Jean-Pierre Rosen and Alfred Strohmeier, editors, *Ada-Europe*, volume 2655 of *Lecture Notes in Computer Science*, pages 368–381. Springer, 2003.

18. Kwok Ping Chan, Tsong Yueh Chen, Dave Towey: Good Random Testing. In Albert Llamosí and Alfred Strohmeier, editors, *Ada-Europe*, volume 3063 of *Lecture Notes in Computer Science*, pages 200–212. Springer, 2004.

19. Kwok Ping Chan, Tsong Yueh Chen, Dave Towey: Restricted Random testing: Adaptive Random Testing by Exclusion. *International Journal of Software Engineering and Knowledge Engineering*, 16(4), 2006 pp. 553–584.

20. Chen, T.Y., Leung, H., Mak, I.K.: Adaptive Random Testing. In Michael, J. Maher, editor, *ASIAN*, volume 3321 of *Lecture Notes in Computer Science*, pages 320–329. Springer, 2004.

21. Chen, T.Y., Leung, H., Mak, I.K., Dave Towey.: Distance-based Adaptive Random Testing. In Preparation.

22. Chen, T.Y., Yu, Y.T.: On the Relationship between Partition and Random Testing. *IEEE Transactions on Software Engineering*, 20(12), Dec 1994, pp. 977–980.

23. Tsong Yueh Chen, Fei-Ching Kuo, Robert G. Merkel, Sebastian P.: Ng. Mirror Adaptive Random Testing. *Information and Software Technology*, 46(15), 2004, pp. 1001–1010.

24. Ilinca, Ciupa, Andreas, Leitner, Manuel Oriol, Bertrand Meyer: Object Distance And Its Application To Adaptive Random Testing Of Object-Oriented Programs. In Johannes Mayer and Robert, G. Merkel, editors, *Random Testing*, 2006, pp. 55–63. ACM.

25. Joe, W., Duran, Simeon Ntafos: A Report on Random Testing. In *ICSE '81: Proceedings of the 5th international conference on Software engineering*, pages 179–183, Piscataway, NJ, USA, 1981. IEEE Press.

26. Joe, W. Duran, Simeon, C.: Ntafos. An Evaluation of Random Testing. *IEEE Transactions on Software Engineering*, 10(4), (1984), pp. 438–444.

27. Elfriede, Dustin: Lessons in Test Automation. *Software Testing & Quality Engineering*, 1(5), 1999 pp. 16–21.

28. Justin, E., Forrester, Barton, P., Miller: An Empirical Study of the Robustness of Windows NT Applications Using Random Testing. In *4th USENIX Windows Systems Symposium, August 2000*, 2000, pp. 59–68.

29. Frankl, P.G., Weyuker, E.J.: A Formal Analysis of the Fault-Detecting Ability of Testing Methods. *IEEE Transactions on Software Engineering*, 19(3), 1993 pp. 202–213.

30. Walter J. Gutjahr. Partition Testing vs. Random Testing: The Influence of Uncertainty. *IEEE Transactions on Software Engineering*, 25(5), 1999, pp. 661–674.

31. Hamlet, D., and Taylor, R.: Partition Testing Does Not Inspire Confidence. *IEEE Transactions on Software Engineering*, 16(12), 1990, pp. 1402–1411.

32. Hamlet, R. Random Testing. In J.Marciniak, editor, *Encyclopedia of Software Engineering*, 1994, pp. 970–978. Wiley.

33. Richard, G. Hamlet. Predicting Dependability by Testing. In *International Symposium on Software Testing and Analysis*, 1996, pp. 84–91.

34. Richard, G. Hamlet, Dave Mason, and Denise M. Woit. Theory of Software Reliability Based on Components. In *International Conference on Software Engineering*, 2001, pp. 361–370.

35. Søren Lauesen and Houman Youuessl. Is Software Quality Visible in the Code? *IEEE Software*, 15(4), 1998, pp. 69–73.

36. Leung, H., Tse, T.H., Chan, F.T., and Chen, T.Y.: Test Case Selection With and Without Replacement. *Inf. Sci. Inf. Comput. Sci*, 129(1-4), 2000, pp. 81–103.

37. Henry Lieberman and Christopher Fry. Will Software Ever Work? *Communications of the ACM*, 44(3), 2001, pp. 122–124.

38. Loo, P.S., and Tsai, W.K.: Random Testing Revisited. *Information and Software Technology*, 30(9), 1988, pp. 402–417.

39. Marick, B.: When Should a Test be Automated. http://www.testing.com/writings/automate.pdf, 1998. [Online; accessed 11-October-2007].

40. Shaul Markovitch and Paul D. Scott. The Role of Forgetting in Learning. In *Proceedings of The Fifth International Conference on Machine Learning*, pages 459–465, Ann Arbor, MI, 1988. Morgan Kaufmann.

41. Tim Menzies and Bojan Cukic. When to Test Less. *IEEE Software*, 17(5), 2000, pp. 107–112.

42. Miller, B.P., Fredriksen, L., So, B.: An Empirical Study of the Reliability of UNIX Utilities. *Communications of the ACM*, 33(12), Dec 1990 .

43. Miller, B.P., Koski, D., Lee, C., Maganty, V., Murthy, R., Natarajan, A. and Steidl, J.: Fuzz Revisited: A Re-examination of the Reliability of UNIX Utilities and Services. Technical Report CS-TR-1995-1268, University of Wisconsin, 1996.

44. Tom Mitchell. *Machine Learning*. McGraw Hill, 1997.

45. Glenford, J. Myers. *Software Reliability: Principles And Practices*. John Wiley & Sons, Inc, New York, NY, USA, 1976.

46. Glenford J. Myers. *Art of Software Testing*. John Wiley & Sons, Inc, New York, NY, USA, 1979.

47. Naur, P., and Randell, B. editors. *Software Engineering: Report on a conference sponsored by the NATO Science Committee, Garmisch, Germany, 7th to 11th October 1968*. Scientific Affairs Division, NATO, 1969.

48. David Owen, Tim Menzies, Mats Heimdahl, and Jimin Gao. Finding Faults Quickly in Formal Models Using Random Search. http://menzies.us/pdf/04fmdebug.pdf, 2004. [Online; accessed 12-July-2008].

49. Roger, S.: Pressman. *Software Engineering: A Practitioner's Approach.* McGraw-Hill Higher Education, 5th edition, 2001.
50. Mary Shaw, David Garlan. *Software Architecture: Perspectives on an Emerging Discipline.* Prentice-Hall, Inc., Upper Saddle River, NJ, USA, 1996.
51. Simons, C.L., Parmee, I.C., Coward, P.D.: 35 Years On: To What Extent Has Software Engineering Design Achieved Its Goals? *IEE Proceedings–Software,* 150(6), 2003, pp. 337–350.
52. Slutz, D.: Massive Stochastic Testing of SQL. In *24th International Conference on Very Large Databases (VLDB 98),* 1998, pp. 618–622.
53. Thayer, R., Lipow, M., Nelson, E.: Software Reliability. 1978.
54. Tsoukalas, M.Z., Duran, J.W., Ntafos, S.C.: On Some Reliability Estimation Problems in Random and Partition Testing. *IEEE Transactions on Software Engineering,* 19(7), 1993, pp. 687–697.
55. Benjamin Tyler, Neelam Soundarajan: Black-Box Testing of Grey-Box Behavior. In *FATES,* 2003, pp. 1–14.
56. Elaine, J. Weyuker, Bingchiang Jeng: Analyzing partition testing strategies, *IEEE Transactions on Software Engineering,* 17(7), 1991, pp. 703–711.
57. Whittaker, J.A.: What is software testing? And why is it so hard?, *IEEE Software,* 17(1), 2000, pp. 70–79.
58. Yoshikawa, T., Shimura, K., and Ozawa, T.: Random program generator for Java JIT compiler test system. In *3rd International Conference on Quality Software (QSIC 2003),* 2003, pp. 20–24.
59. Zadeh, L.A.: Fuzzy: *Information and Control,* 3(8), 1965, pp. 338–353.

Chapter 4

TRANSPARENT SHAPING: A METHODOLOGY FOR ADDING ADAPTIVE BEHAVIOR TO EXISTING SOFTWARE SYSTEMS AND APPLICATIONS

S. MASOUD SADJADI

School of Computing and Information Sciences
Florida International University
Miami, Florida 33199, USA
sadjadi@cs.fiu.edu

PHILIP K. MCKINLEY and BETTY H.C. CHENG*

Department of Computer Science and Engineering
Michigan State University
East Lansing, Michigan 48824, USA
{mckinley,chengb}@cse.msu.edu

The need for adaptability in software is growing, driven in part by the emergence of pervasive and autonomic computing. In many cases, it is desirable to enhance existing programs with adaptive behavior, enabling them to execute effectively in dynamic environments. In this chapter, we introduce an innovative software engineering methodology called *transparent shaping* that enables dynamic addition of adaptive behavior to existing software systems and applications. We describe an approach to implementing transparent shaping that combines four key software development techniques: aspect-oriented programming to realize separation of concerns at development time, behavioral reflection to support software reconfiguration at run time, component-based design to facilitate independent development and deployment of adaptive code, and adaptive middleware to encapsulate the adaptive functionality. After presenting the general methodology, we discuss two specific realizations of transparent shaping that we have developed and used to create adaptable systems and applications from existing ones.

1. Introduction

A software application is *adaptable* if it can change its behavior dynamically (at run time) in response to transient changes in its execution

---

*Corresponding author.

environment or to permanent changes in its requirements. Recent interest in designing adaptable software is driven in part by the emergence of pervasive computing and the demand for autonomic computing.[1] *Pervasive computing* promises anywhere, any time access to data and computing resources with few limitations and disruptions.[2] The need for adaptability in pervasive computing is particularly evident at the "wireless edge" of the Internet, where software in mobile devices must balance conflicting concerns such as quality-of-service (QoS) and energy consumption when responding to variability of conditions (e.g., wireless network loss rate). *Autonomic computing*[3] refers to self-managed, and potentially self-healing, systems that require only high-level human guidance. Autonomic computing is critical to managing the myriad of sensors and other small devices at the wireless edge, but also in managing large-scale computing centers and protecting critical infrastructure (e.g., financial networks, transportation systems, power grids) from hardware component failures, network outages, and security attacks.

Developing and maintaining adaptable software are nontrivial tasks. An adaptable application comprises *functional* code that implements the business logic of the application and supports its imperative behavior, and *adaptive* code that implements the adaptation logic of the application and supports its adaptive behavior. The difficulty in developing and maintaining adaptable applications is largely due to an inherent property of the adaptive code, that is, the adaptive code tends to *crosscut* the functional code. Example crosscutting concerns include QoS, mobility, fault tolerance, recovery, security, self auditing, and energy consumption. Even more challenging than developing new adaptable applications is enhancing *existing* applications, such that they execute effectively in new, dynamic environments not envisioned during their design and development. For example, many non-adaptive applications are being ported to mobile computing environments, where they require dynamic adaptation.

This chapter describes a new software engineering methodology, called *transparent shaping*, that supports the design and development of adaptable programs from existing programs without the need to modify the existing programs' source code directly. We argue that *automatic* generation of an adaptable program from a non-adaptable one is important to maintaining program integrity, not only because it avoids errors introduced by manual changes, but because it provides traceability for the adaptations and enables the program to revert back to its original behavior if necessary. Our approach to implementing transparent shaping combines four key technologies: *aspect-oriented programming* to enable separation of concerns at development time, *behavioral reflection* to enable software reconfiguration at run time, *component-based design* to enable independent development and deployment of adaptive code, and *adaptive middleware* to help insulate application code from adaptive functionality. To demonstrate the effectiveness of this approach, we describe two realizations of transparent shaping that we have developed and used to create adaptable applications.

The remainder of this chapter is organized as follows. Section 2 discusses the four main components of our approach. Section 3 provides an overview of transparent shaping and describes its relationship to program families.[4] Sections 4 and 5, respectively, describe two realizations of transparent shaping; one is middleware-based and the other is language-based. Section 6 discusses how transparent shaping complements other research in adaptive software. Section 7 presents conclusions and identifies several directions for future research.

2. Basic Elements

Transparent shaping integrates four key technologies: separation of concerns, behavioral reflection, software components, and middleware. In this section, we briefly review each technology and its role in transparent shaping.

Separation of concerns[5] enables the separate development of the functional code from the adaptive code of an application. This separation simplifies development and maintenance, while promoting software reuse. Moreover, since adaptation often involves crosscutting concerns, this separation also facilitates transparent shaping. In our approach, we use aspect-oriented programming (AOP),[6,7] an increasingly common approach to implementing separation of concerns in software. While object-oriented programming introduces abstractions to capture commonalities among classes in an inheritance tree, crosscutting concerns are scattered among different classes, thus complicating the development and maintenance of applications. Conversely, in AOP the code implementing such crosscutting concerns, called *aspects*, is developed separately from other parts of the system. Later, for example during compilation, an *aspect weaver* can be used to weave different aspects of the program together to form a program with new behavior. Predefined locations in the program where aspect code can be woven are called *pointcuts*.

In traditional AOP, after compilation the aspects are tangled (via weaving) with the functional code. To facilitate dynamic reconfiguration, transparent shaping needs a way to enable separation of concerns to persist into run time. This separation can be accomplished using *behavioral reflection*,[8] the second key technology for transparent shaping. Behavioral reflection enables a system to "open up" its implementation details at run time.[9] A reflective system has a self representation that deals with the computational aspects (implementation) of the system, and is *causally connected* to the system. The self-representation of a reflective system is realized by metaobjects residing in the metalevel, which is separated from the actual system represented by objects in the base level. A *metaobject* is an entity that manipulates, creates, describes, or implements other objects, which in turn called *base objects*, and might store some information about these base objects such as their type, interface, class, methods, attributes, variables, functions, and control structures. By incorporating crosscutting concerns associated with the system as part of its self-representation, the resulting code at run time is not tangled and therefore can be reconfigured dynamically. When combined with AOP, behavioral reflection enables dynamic weaving of crosscutting concerns into an application at run time.[10]

The third major technology that supports transparent shaping is component-based design. *Software components* are software units that can be independently developed, deployed, and composed by third parties.[11] Well-defined interface specifications supported in component-based design enable adaptive code to be developed independently from the functional code, and potentially by different parties, using the interface as a contract. Component-based design supports two types of composition. In static composition, a developer can combine several components at compile time to produce an application. In dynamic composition, the developer can add, remove, or reconfigure components within an application at run time. When combined with behavioral reflection, component-based design enables a "plug-and-play" capability for adaptive code to be incorporated with functional code at run time that facilitates development and maintenance of adaptable software.

Finally, in many cases it is desirable to hide the adaptive behavior from the application using middleware. Traditionally, *middleware* is intended to mask the distribution of resources across a network and hide differences among computing platforms and networks.[12] As observed by several researchers,[13] however, middleware is also an ideal place to incorporate adaptive behavior for many different crosscutting concerns. *Adaptive middleware* enables dynamic reconfiguration of middleware services while an application is running, adjusting the middleware behavior to environmental changes dynamically. Our approach to transparent shaping uses adaptive middleware in two ways. In the first, transparent shaping adds adaptive behavior to a middleware platform already supporting the application. In the second, transparent shaping is used to weave calls to adaptive middleware into an application.

3. General Approach

By generating adaptable programs from existing ones, transparent shaping is intended to support the reuse of those applications in environments whose characteristics were not necessarily anticipated during the original design and development. Therefore, a challenge in transparent shaping is finding a way to produce adaptable programs that share the business logic of the original program and differ only in the new adaptive behavior.

As illustrated in Fig. 1, one way to formulate this problem is using *program families*, a well-established concept in the software engineering community. A program family[4] is a set of programs whose extensive commonalities justify the expensive effort required to study and develop them as a whole, rather than individually. In short, transparent shaping can be viewed as a methodology that produces a family of adaptable programs from an existing non-adaptable program. The adaptable program comprises the original program code that remains fixed during program execution, and adaptive code that can be replaced with other adaptive

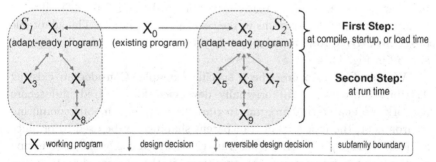

Fig. 1. A transparent shaping design tree illustrating a family of adaptable programs produced from an existing program, which is the root of this tree. Children of the root are adapt-ready programs. Other descendants are adaptable programs.

code dynamically. Replacing one piece of adaptive code with another piece of adaptive code converts an adaptable program into another adaptable program in the corresponding family. This conversion is possible in this programming model, because the adaptive code is not tangled with the functional code. We use the term *composer* to refer to the entity that performs this conversion. The composer might be a human — a software developer or an administrator interacting with a running program through a graphical user interface — or a piece of software — a dynamic aspect weaver, a component loader, a run-time system, or a metaobject.

Transparent shaping produces adaptable programs in two steps. In the first step, an *adapt-ready* program[14] is produced at compile, startup, or load time using static transformation techniques. An adapt-ready program is a program whose behavior is initially equivalent to the original program, but which can be adapted at run time by insertion or removal of adaptive code at certain points in the execution path of the program, called *sensitive joinpoints*. To support such operations, the first step of transparent shaping weaves interceptors, referred to as *hooks*, at the sensitive joinpoints, which may reside inside the program code itself, inside its supporting middleware, or inside the system platform. Example techniques for implementing hooks include aspects (compile time), CORBA portable interceptors[15] (startup time), and byte-code rewriting[16] (load time).

In the second step, executed at run time, the hooks in the adapt-ready program are used by the composer to convert the adapt-ready program into an adaptable program in the corresponding subfamily, as conditions warrant. Adapt-ready programs derived from the same existing program differ in their corresponding sensitive joinpoints and hooks. We note that the available hooks in an adapt-ready program limit its dynamic behavior. In other words, each adapt-ready program can be converted to a limited number of adaptable programs in the corresponding family. The adaptable programs derived from an adapt-ready program form a subfamily (e.g., $S1$ and $S2$ in Fig. 1).

We use Fig. 1 to describe a specific example. Consider an existing distributed program (X_0) originally developed for a wired and secure network. To enable this program to run efficiently in a mobile computing environment, the first step of transparent shaping can be used to produce an adapt-ready version of this program (X_1), which has hooks intercepting all the remote interactions. At run time, if the system detects a low quality wireless connection, the composer can insert adaptive code for tolerating long periods of disconnection into the adapt-ready program

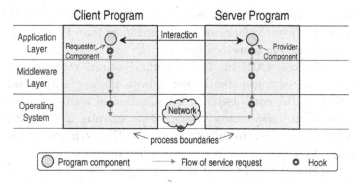

Fig. 2.   Alternative places to insert hooks.

(producing X_4 from X_1). Later, if the user enters an insecure wireless network, the composer can insert adaptive code for encryption/decryption of the remote interactions into the program (producing X_8 from X_4). Finally, when the user returns to an area with a secure and reliable wireless connection, the composer can remove the adaptive code for both security and connection-management to avoid unnecessary performance overhead due to the adaptive code (producing X_4 from X_8 and X_1 from X_4, respectively).

We identify three approaches to realize transparent shaping that differ according to the placement of hooks (see Fig. 2): (1) hooks can be incorporated inside an application program itself, (2) inside its supporting middleware, or (3) inside the system platform (operating system and network protocols). A number of projects on cross-layer adaptation use the third approach.[17-19] In this paper, we consider only the first two methods, where the hooks are incorporated either inside the middleware or inside the application. Next, we describe two concrete realizations of each type of transparent shaping. The first, described in Section 4, is a middleware-based approach that uses CORBA portable interceptors[15] as hooks. The second, described in Section 5, uses a combination of aspect weaving and metaobject protocols to introduce dynamic adaptation to the application code directly. Both realizations adhere to the general model described above.

4. Middleware-Based Transparent Shaping

The first realization of transparent shaping we describe is the *Adaptive CORBA Template (ACT)*,[20,21] which we developed to enable dynamic adaptation in existing CORBA programs. CORBA was one of the first widely used middleware platforms introduced more than 17 years ago. It is still commonly used in numerous systems.

ACT enhances CORBA to support dynamic reconfiguration of middleware services transparently, not only to the application code, but also to the CORBA code itself. As a realization of transparent shaping, ACT produces an adapt-ready version of an existing CORBA program by introducing a hook to intercept all CORBA remote interactions. Specifically, ACT uses CORBA portable interceptors,[15] which can be incorporated into a CORBA program at startup time using a command-line parameter. Later at run time, these hooks can be used to insert adaptive code into the adapt-ready program, which in turn can adapt the requests, replies, and exceptions passing through the CORBA Object Request Brokers (ORBs). In this manner, ACT enables run-time improvements to the program in response to unanticipated changes in its execution environment, effectively producing other members of the adaptable program family dynamically.

4.1. ACT Architectural Overview

Figure 3 shows the flow of a request/reply sequence in a simple CORBA application using ACT. For clarity, CORBA ORB details such as stubs and skeletons are not shown. ACT comprises two main components: a

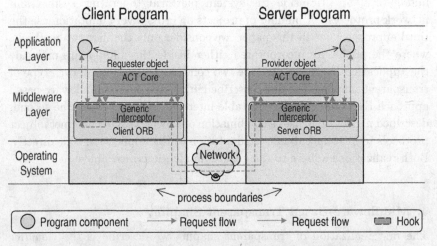

Fig. 3.   ACT configuration in the context of a simple CORBA application.

generic interceptor and an ACT core. A *generic interceptor* is a specialized request interceptor that is registered with the ORB of a CORBA application at startup time. The *client* generic interceptor intercepts all outgoing requests and incoming replies (or exceptions) and forwards them to its ACT core. Similarly, the *server* generic interceptor intercepts all the incoming requests and outgoing replies (or exceptions) and forwards them to its ACT core. A CORBA application is called *adapt-ready* if a generic interceptor is registered with all its ORBs at startup time. If, in addition to the generic interceptors, all the ACT core components are also loaded into the application, the application is called *ACT-ready*. Making the application ACT-ready can be done either at startup time or at run time.

4.2. *ACT Core Components*

Figure 4 shows the flow of a request/reply sequence intercepted by the client ACT core. The components of the core include dynamic interceptors, a proxy, a decision maker, and an event mediator. Each component is described in turn.

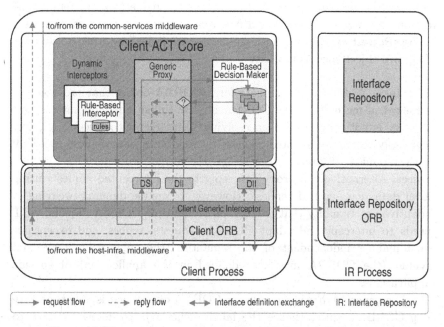

Fig. 4.   ACT core components interacting with the rest of the system.

Dynamic Interceptors. According to the CORBA specification,[15] a request interceptor is required to be registered with an ORB at the ORB initialization time. The ACT core enables registration of request interceptors after the ORB initialization time (at run time) by publishing a CORBA interceptor-registration service. Such request interceptors are called *dynamic interceptors*. Dynamic interceptors can be unregistered with the ORB at run time also. In contrast, a request interceptor that is registered with the ORB at startup time is called a *static interceptor* and cannot be unregistered with the ORB during run time. We note that the code developed for a static interceptor and that for a dynamic interceptor can be identical, the difference being the time at which they are registered. In ACT, only generic interceptors are static.

A *rule-based interceptor* is a particular type of dynamic interceptor that uses a set of rules to direct the operations on intercepted requests. The rules can be inserted, removed, and modified at run time. A *rule* consists of two objects: a condition and an action. To determine whether a rule matches a request, a rule-based interceptor consults its condition object. Once a match is found, the interceptor sends the request to the action object of the rule. Since it is part of a CORBA portable interceptor, the action object cannot itself reply to the request or modify the request parameters.[15] The action object can, however, send new requests, record statistics, or raise a ForwardRequest exception, causing the request to be forwarded to another CORBA object such as a proxy.

Proxies. A *proxy* is a surrogate for a CORBA object that provides the same set of methods as the CORBA object. Unlike a request interceptor, a proxy is not prohibited from replying to intercepted requests. A proxy can reply to the intercepted request by sending a new request (possibly with modified arguments) to either the target object or to another object. Alternatively, a proxy can reply to the intercepted requests using local data (e.g., cached replies). However, to enable dynamic weaving of adaptive functionality that is common to multiple CORBA objects, ACT needs to intercept and adapt CORBA requests, replies, and exceptions in a manner independent of the semantics (the application logic) and syntax (the CORBA interfaces defined in the application) of specific applications.

The *generic proxy* is a particular CORBA object that is able to receive *any* CORBA request (hence the label "generic"). To determine how to handle a particular request, the generic proxy accesses the CORBA interface repository,[15] which provides all the IDL descriptions for CORBA requests. The repository executes as a separate process and is usually accessed through the ORB. Most CORBA ORBs provide a configuration file or support a command-line argument that enables the user to introduce the interface repository to the application ORB. Providing IDL information to the generic proxy in this manner implies no need to modify or recompile the application source code. The interface repository, however, requires access to the CORBA IDL files used in the application.

In default operation mode, the generic proxy intercepts CORBA requests, acquires the request specifications from a CORBA interface repository, creates similar CORBA requests and sends them to the original targets, and forwards replies from those targets back to the original clients. A generic proxy also publishes a CORBA service that can be used to register a *decision maker*.

Decision Makers. A *decision maker* assists proxies in replying to intercepted requests as depicted in Fig. 4. A decision maker receives requests from a proxy and, similar to a rule-based interceptor, uses a set of rules to direct the operation on the intercepted requests. However, unlike a rule-based interceptor, a decision maker is not prohibited from replying to the requests.

4.3. *ACT Operation*

In addition to showing the ACT core components, Fig. 4 also illustrates the sequence of a request/reply inside the ACT core, which contains a rule-based interceptor, a generic proxy, and a rule-based decision maker. First, a request from the client application is intercepted by the rule-based interceptor, which checks its rules for possible matches. A default rule, initially inserted in its knowledge base, directs the rule-based interceptor to raise a ForwardRequest exception, which results in its forwarding the request to the generic proxy. When the generic proxy receives the request, it acquires the request interface definition via the application ORB, which in turn retrieves the information from the interface repository. The proxy creates a new request and forwards it to the rule-based decision maker. The rule-based decision maker checks its knowledge base for possible matches to the request. Depending on the implementation of the rules, the decision maker may return either a modified request to the generic proxy or a reply to the request. If the decision maker returns the request (or a modified request), then the generic proxy will continue its operation by invoking the request. If the reply to the request is returned by the decision maker, then the proxy replies to the original request using the reply from the decision maker. The generic proxy uses the CORBA dynamic skeleton interface (DSI)[15] to receive any type of request. The generic proxy and the rule-based decision maker use the CORBA dynamic invocation interface (DII)[15] to create and invoke a new request dynamically.

4.4. ACT/J Implementation

We have developed an instance of ACT in Java, called *ACT/J*, to evaluate ACT in practice. ACT/J was tested over ORBacus,[22] a CORBA-compliant ORB distributed by IONA Technologies. ORBacus,[22] like JacORB,[23] TAO,[24] and many other CORBA ORBs, supports *CORBA portable interceptors*,[15] which is the only requirement for using ACT.

To make a CORBA application ACT-ready at the application startup time, we need to resolve the following bootstrapping issues. First, we need to register a generic interceptor with the application ORB. Like many other ORBs, ORBacus uses a configuration file that enables an administrator to register a CORBA portable interceptor with the application ORB. JacORB and TAO use a similar approach. Second, since the components in the ACT core are also CORBA objects, they require an ORB to support their operation (registration of services, and so on). Therefore, we need either to obtain a reference to the application ORB for this purpose, or to create a new ORB. ORBacus does provide such a reference, although the CORBA specification does not support this feature. To implement ACT/J over an ORB that does not provide such a reference, we simply create a new ORB, although its use introduces additional overhead.

To test the operation of ACT/J, we developed two administrative consoles: the Interceptor Registration Console and the Rule Management Console. Please note that in this study the composer is assumed to be a human, who performs dynamic adaptation using the administrative consoles. The *Interceptor Registration Console* enables a user to manually register a dynamic interceptor. This console first obtains a generic interceptor name from the user and checks if the generic interceptor is registered with the CORBA naming service. Next, the user can register a dynamic interceptor with the generic interceptor. The *Rule Management Console* allows a user to manually insert rules into rule-based interceptors.

4.5. *ACT/J Case Study*

To evaluate the effectiveness of ACT/J to support self-management in existing CORBA applications, without modifying the application code, we conducted a case study in which self-optimization is enabled in an existing application. Additional experiments involving IP handoff, are described in an accompanying technical report.[25] We begin with a brief overview of the application and the experimental environment, followed by the description of the experiment. The experiment shows how ACT/J could be used to support autonomic computing in either a generic or application-specific manner.

For the application, we adopted an existing distributed image retrieval application developed by BBN Technologies.[26] The application has two parts, a client that requests and displays images, and a server that stores the images and replies to requests for them. In this study, we treat the application as though it were used for surveillance, with a mobile user executing the client code on a laptop and monitoring a physical facility through continuous still images from multiple camera sources. For the experiment described later in this section, we executed the server on a desktop computer connected to a 100 Mbps wired network and the client on a laptop computer connected to a three-cell 802.11b wireless network. Both the desktop and laptop systems are running the Linux operating system.

Figure 5 shows the physical configuration of the three access points used in the experiment. (The wireless cells are drawn as circles for simplicity — the actual cell shapes are irregular, due to the physical construction of the building and orientation of antennas.) AP-1 and AP-3 provide 11 Mbps connections, whereas AP-2 provides only 2 Mbps. The desktop running the server application is close to AP-1. AP-1 and AP-2 are managed by our Computer Science and Engineering Department, whereas AP-3 is managed by the College of Engineering. This difference implies that the IP address assigned to the client laptop needs to change as the user moves from a CSE wireless cell to a College cell. The server provides four different versions of each image, varying in size and quality. Typical comparative file sizes are 90KB, 25KB, 14KB, and 4KB.

To investigate how ACT/J can support self-management, we developed an application-specific rule that maintains the frame rate of the application by controlling the image size or inserting inter-frame delays dynamically. The original image retrieval application operates in a default mode, which retrieves and plays images as fast as possible. ACT/J enables a developer

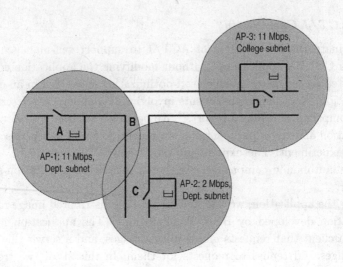

Fig. 5.   The configuration of the access points used in the experiment.

to weave the rule into the application at run time, thereby providing new functionality (frame rate control) transparently with respect to the application. The self-optimization rule maintains the frame rate of the application in the presence of dynamic changes to the wireless network loss rate, the network (wired/wireless) traffic, and CPU availability.

We developed a user interface, called the Automatic Adaptation Console, which displays the application status and also enables the user to enter quality-of-service preferences (see Fig. 6). The rule uses several parameters to decide on when and how to adapt the application in order to maintain the frame rate. These parameters have default values as shown in the figure, but can be modified at run time by the user. The Average Frame Rate Period indicates the period during which the average frame rate should be calculated to be considered for adaptation. The Stabilizing Period specifies the amount of time that the rule should wait until the last adaptation stabilizes; also if a sudden change occurs in the environment such as hand-off from one wireless cell to another one, then the system should wait for this period before it decides on the stability of the system. The rule detects a stable situation using the Acceptable Rate Deviation; when the frame rate deviation goes below this value, the system is considered stable. Similarly, the rule detects an unstable situation, if the instantaneous frame rate deviation goes beyond the Unacceptable Rate Deviation value. The rule also maintains a history of the round-trip delay

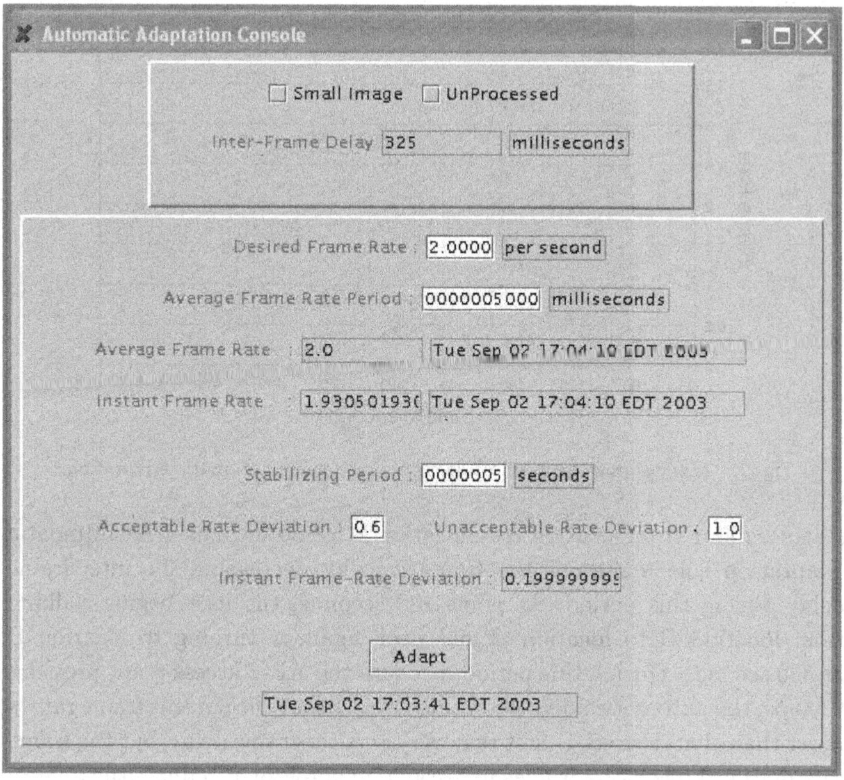

Fig. 6. Automatic Adaptation Console.

associated with each request in each wireless cell. Using this history and the above parameters, the rule can decide to maintain the frame rate either by increasing/decreasing the inter-frame delay or by changing the request to ask for a different version of the image with smaller/larger size. The default behavior of the rule is to display images that are as large as possible, given the constraints of the environment.

Figure 7 shows a trace demonstrating automatic adaptation of the application in the following scenario. In this experiment, the user has selected a desired frame rate of 2 frames per second, as shown in Fig. 6. For the first 60 seconds of the experiment, the user stays close to the location A (Fig. 5). The rule detects that the desired frame rate is lower than the maximum possible frame rate, based on observed round-trip times. Hence, it inserts an inter-frame delay of approximately 200 milliseconds to maintain the frame rate at about 2 frames per second. At point 120 seconds, the user

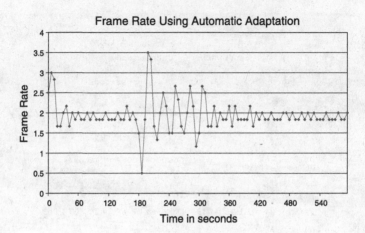

Fig. 7.   Maintaining the application frame rate using automatic adaptation.

starts walking from location A to location B for 60 seconds. The automatic adaptation rule maintains the frame rate by decreasing the inter-frame delay during this period. At point 180 seconds, the user begins walking from location B to location C and back again, returning to location B at 360 seconds. During this period, because the AP-2 access point provides 2 Mbps, the automatic adaptation rule detects that the current frame rate is lower than that desired. It first removes the inter-frame delay, but the frame rate does not reach to 2 frames per second. Therefore, it reduces the quality of the image by asking for a smaller image size. Now the frame increases beyond that desired, so the automatic adaptation rule inserts an inter-frame delay of 400 milliseconds to maintain the frame rate at 2 frames per second. Although there is some oscillation, the rate stabilizes by time 360 seconds. At this point, the user continues walking from location B to location A, prompting the rule to reverse the actions. First the inter-frame delay is increased to maintain the frame rate, followed by an increase in image size. In this manner, the rule brings the application back to its original behavior. Again, because the current frame rate is higher than expected, an inter-frame delay of about 200 milliseconds is inserted to maintain the frame rate at 2 frames per second.

This result is promising and demonstrates that it is possible to add self-management behavior to an application transparently to the application code. Moreover, the use of a generic proxy enables self-optimization functionality, both application-independent and application-specific, to be added to the application, even at run time.

As a middleware-based realization of transparent shaping, ACT can be used to produce families of adaptable programs from existing CORBA programs, without the need to modify or recompile their source code. Using the generic interceptor as a hook inside middleware at startup time, ACT enables independent development and deployment of adaptive code from the application code at run time. In ACT, adaptive code are realized as software components (rules and proxies) that can be deployed inside the ACT core dynamically. By allowing dynamic insertion and removal of such adaptive code, ACT enables dynamic conversion of an adapt-ready CORBA program to different adaptable programs in its corresponding program subfamily.

5. Language-Based Transparent Shaping

Although transparent shaping can be realized by incorporating hooks inside middleware, as in ACT, many programs do not use middleware explicitly. In this section, we introduce TRAP (Transparent Reflective Aspect Programming),[27] a language-based realization of transparent shaping that supports dynamic adaptation in existing programs developed in class-based, object-oriented programming languages. TRAP uses generative techniques to create an adapt-ready application, without requiring any direct modifications to the existing programs.

With TRAP, the developer selects at compile time a subset of classes in the existing program that are to be reflective at run time. We say a class is *reflective* at run time if its behavior (e.g., the implementation of its methods) can be inspected and modified dynamically. Since many object-oriented languages, such as Java and C++, do not support such functionality inherently, TRAP uses generative techniques to produce an adapt-ready program with hooks that provide the reflective facilities for the selected classes. As the adapt-ready program executes, new behavior can be introduced to the program by insertion and removal of adaptive code via interfaces to the reflective classes.

5.1. *TRAP/J Architectural Overview*

We developed TRAP/J, a prototype instantiation of TRAP for Java programs.[27] The operation of the first step, converting an existing Java program into an adapt-ready program, is depicted in Fig. 8. We assume that the .java source files of the original application are not available. The

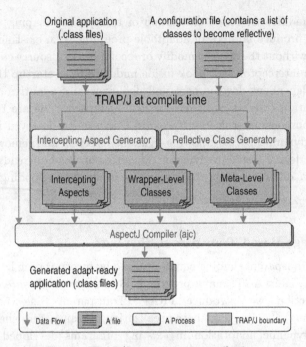

Fig. 8.   TRAP/J operation at compile time.

compiled class files (.class files) of the application and a configuration file containing a list of class names (the ones selected to be reflective) are input to an Aspect Generator and a Reflective Class Generator. For each class name in the list, these generators produce one aspect, one wrapper-level class, and one metalevel class. Next, the generated aspects and reflective classes, along with the original application compiled class files, are passed to the AspectJ compiler (ajc),[28] which weaves the generated and original application code together to produce an adapt-ready application. The second step occurs at run time, when new behavior can be introduced to the adapt-ready application using the wrapper- and meta-level classes (also referred to as the adaptation infrastructure). Specifically, the interface of the metalevel class includes services that enable methods of the wrapper-level class to be overridden at run time with new implementations, called *delegates*.

Figure 9 illustrates the interaction among the Java Virtual Machine (JVM) and the administrative consoles (GUI). First, the adapt-ready application is loaded by the JVM. At the time each metaobject is instantiated, it registers itself with the Java rmiregistry using a unique

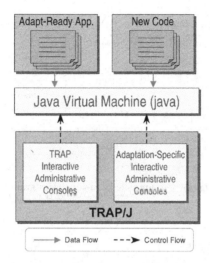

Fig. 9. TRAP/J run-time support.

ID. Next, if an adaptation is required, the composer dynamically adds new code to the adapt-ready application at run time, using Java RMI to interact with the metaobjects. As part of the behavioral reflection provided in the adaptation infrastructure, a metaobject protocol (MOP) is supported in TRAP/J that allows interception and reification of method invocations targeted to objects of the classes selected at compile time to be adaptable.

5.2. *TRAP/J Run-Time Model*

To illustrate the operation of TRAP/J, let us consider a simple application comprising two classes, Service and Client, and three objects, (client, s1, and s2). Figure 10 depicts a simple run-time class graph for this application that is compliant with the run-time architecture of most class-based object-oriented languages. The class library contains Service and Client classes, and the heap contains client, s1, and s2 objects. The "instantiates" relationship among objects and their classes are shown using dashed arrows, and the "uses" relationships among objects are depicted with solid arrows.

Figure 11 illustrates a layered run-time class graph model for this application. Please note that the base-level layer depicted in Fig. 11 is equivalent to the class graph illustrated in Fig. 10. For simplicity, only the "uses" relationships are represented in Fig. 11. The wrapper level contains the generated wrapper classes for the selected subset of base-level classes

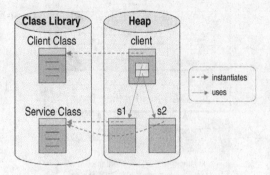

Fig. 10.   A simplified run-time class graph.

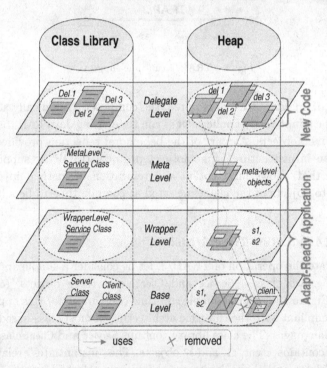

Fig. 11.   TRAP layered run-time model.

and their corresponding instances. The base-level client objects use these wrapper-level instances instead of base-level service objects. As shown, s1 and s2 no longer refer to objects of the type Service, but instead refer to objects of type ServiceWrapper class. The metalevel contains the generated metalevel classes corresponding to each selected base-level class and their corresponding instances. Each wrapper class has exactly one associated metalevel class, and associated with each wrapper object can be at most one metaobject. In this way, the behavior of each object in response to each message is dynamically programmable, using the generic method execution MOP provided in TRAP/J.

Finally, the delegate level contains adaptive code that can dynamically override base-level methods that are wrapped by the wrapper classes. Adaptive code is introduced into TRAP/J using *delegate* classes. A delegate class can contain implementation for an arbitrary collection of base-level methods of the wrapped classes, enabling the localization of a crosscutting concern in a delegate class. A composer can program metaobjects dynamically to redirect messages destined originally to base-level methods to their corresponding implementations in delegate classes. Each metaobject can use one or more delegate instances, enabling different crosscutting concerns to be handled by different delegate instances. Moreover, delegates can be shared among different metaobjects, effectively providing a means to support dynamic aspects.

For example, let us assume that we want to adapt the behavior of a socket object (instantiated from a Java socket class such as the Java.net.MulticastSocket class) in an existing Java program at run time. First, at compile time, we use TRAP/J generators to generate the wrapper and metaobject classes associated with the socket class. Next, at run time, a composer can program the metaobject associated with the socket object to support dynamic reconfiguration. Programming the metaobject can be done by introducing a delegate class to the metaobject at run time. The metaobject then loads the delegate class, instantiates an object of the delegate class, intercepts all subsequent messages originally targeted to the socket object, and forwards the intercepted messages to the delegate object. Let us assume that the delegate object provides a new implementation for the **send()** method of the socket class. In this case, all subsequent messages to the **send()** method are handled by the delegate object and the other messages are handled by the original socket object. Alternatively, the delegate object could modify the intercepted messages and then forward them back to the socket object, resulting in a new behavior. TRAP/J allows the composer to remove delegates at run time, bringing the object behavior back to its original implementation. Thus, TRAP/J is a non-invasive[29] approach to dynamic adaptation.

In an earlier study,[27] we developed a delegate that effectively allows selected Java sockets in an existing program to be replaced with adaptable communication middleware components called MetaSockets. A *MetaSocket* is created from existing Java socket classes, but its structure and behavior can be adapted at run time in response to external stimuli such as dynamic wireless channel conditions. Specifically, data sent or received on the socket is passed through a pipeline of filters. A MetaSocket itself can be reconfigured dynamically in its filter pipeline. The filter pipeline can be reconfigured dynamically, that is, filters can be inserted and removed, in response to changes in changing conditions. Moreover, the filter components can be developed by third parties and can be independent of the functional code of an application. Using TRAP/J and MetaSockets, we demonstrated how to transform existing network applications into adaptive applications that can better tolerate dynamic conditions on wireless networks.[27]

5.3. *TRAP/J Case Study*

To demonstrate how TRAP/J can be used to produce adaptable programs from an existing program without the need to modify the existing program source code directly, we use the Audio Streaming Application, called ASA, that is designed to stream interactive audio from a microphone at one network node to multiple receiving nodes. The original application was developed for wired networks. Our goal is to adapt this application to wireless environments, where the packet loss rate is dynamic and location dependent.

In this case study, we configured the experiments in an *ad hoc* wireless network as illustrated in Fig. 12. A laptop workstation transmits an audio stream to multiple wireless iPAQs over an 802.11b (11 Mbps) *ad hoc* wireless local area network (WLAN). Please note that unlike in wired networks, in wireless networks factors such as signal strength, interference,

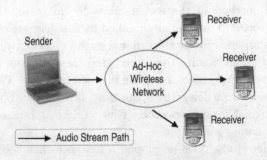

Fig. 12.   Audio streaming in a wireless LAN.

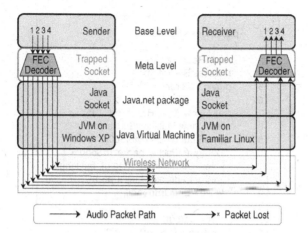

Fig. 13. Adaptation strategy.

and antenna alignment produce dynamic and location-dependent packet losses. In current WLANs, these problems affect multicast connections more than unicast connections, since the 802.11b MAC layer does not provide link-level acknowledgements for multicast frames.

Figure 13 illustrates the strategy we used to enable ASA to adapt to variable channel conditions in wireless networks. However, we used TRAP/J to modify ASA *transparently* such that it uses MetaSockets instead of Java multicast sockets. The particular MetaSocket adaptation used here is the dynamic insertion and removal of *forward-error correction* (FEC) filters.[30] Specifically, an FEC encoder filter can be inserted and removed dynamically at the sending MetaSocket, in synchronization with an FEC decoder being inserted and removed at each receiving MetaSocket. Use of FEC under high packet loss conditions reduces the packet loss rate as observed by the application. Under low packet loss conditions, however, FEC should be removed so as not to waste bandwidth on redundant data.

Making ASA Adapt-Ready. Figure 14 shows excerpted code for the original Sender class. The main method creates a new instance of the Sender class and calls its run method. The run method first creates an instance of AudioRecorder and MulticastSocket and assigns them to the instance variables, ar and ms, respectively. The multicast socket (ms) is used to send the audio datagram packets to the receiver applications. Next, the run method executes an infinite loop that, for each iteration, reads live audio data and transmits the data via the multicast socket.

```
1    public class Sender
2    {
3      AudioRecorder ar;
4      MulticastSocket ms;
5      public void run()
6      { ...
7        ar = new AudioRecorder(...);
8        ms = new MulticastSocket();
9        byte[] buf = new byte[500];
10       DatagramPacket packetToSend =
11         new DatagramPacket(buf, buf.length,
12         target_address, target_port);
13       while (!EndOfStream)
14       {
15         ar.read(buf, 0, 500);
16         ms.send(packetToSend);
17       } // end while ...
18     }
19   } // end Sender
```

Fig. 14.   Excerpted code for the Sender class.

Compile-Time Actions.   The Sender.java file and a file containing only the java.net.MulticastSocket class name are input to the TRAP/J aspect and reflective generators. The TRAP/J class generators produce one aspect file, named Absorbing_MulticastSocket.aj (for base-level), and two reflective classes, named WrapperLevel_MulticastSocket.java (wrapper level) and MetaLevel_MulticastSocket.java (metalevel). Next, the generated files and the original application code are compiled using the AspectJ compiler (ajc) to produce the adapt-ready program. We note that if ajc could accept .class files instead of .java files, then we would not even need the original source code in order to make the application adapt-ready.

Generated Aspect.   The aspect generated by TRAP/J defines an initialization pointcut and the corresponding around advice for each public constructor of the MulticastSocket class. An around advice causes an instance of the generated wrapper class, instead of an instance of MulticastSocket, to serve the sender. Figure 15 shows excerpted code for the generated Absorbing_MulticastSocket aspect. This figure shows the "initialization" pointcut (lines 3–4) and its corresponding advice (lines 6–11) for the MulticastSocket constructor used in the Sender class. Referring back to the layered class graph in Fig. 11, the sender (client) uses an instance

```
1    public aspect Absorbing_MulticastSocket
2    {
3      pointcut MulticastSocket() :
4        call(java.net.MulticastSocket.new()) && ...;
5
6      java.net.MulticastSocket around()
7        throws java.net.SocketException
8        : MulticastSocket()
9      {
10       return new WrapperLevel_MulticastSocket();
11     }
12
13     pointcut MulticastSocket_int(int p0) :
14       call(java.net.MulticastSocket.new(int))
15         && args(p0) && ...;
16
17     // Pointcuts and advices around the  nal public methods
18     pointcut getClass(WrapperLevel_MulticastSocket
19       targetObj) :
20       ...;
21   }
```

Fig. 15.   Excerpted generated aspect code.

of the wrapper class instead of the base class. In addition to handling public constructors, TRAP/J also defines a pointcut and an around advice to intercept all public final and public static methods.

Generated Wrapper-Level Class.   Figure 16 shows excerpted code for the WrapperLevel_MulticastSocket class, the generated wrapper class for the MulticastSocket. This wrapper class extends the MulticastSocket class. All the public constructors are overridden by passing the parameters to the super class (base-level class) (lines 4–6). Also, all the public instance methods are overridden (lines 8–29).

To better explain how the generated code works, we step through the details of how the send method is overridden, as shown in Fig. 16. The generated send method first checks whether the metaObject variable, referring to the metaobject corresponding to this wrapper-level object, is null (lines 11–12). If so, then the base-level (super) method is called, as if the base-level method had been invoked directly by another object, such as an instance of sender. Otherwise, a message containing the context information is dynamically created using Java reflection and passed to the metaobject (metaObject) (lines 14–28). It might be the case that a metaobject may need

```
1    public class WrapperLevel_MulticastSocket extends
2    MulticastSocket implements WrapperLevel_Interface {
3
4    // Overriding the base-level constructors.
5    public WrapperLevel_MulticastSocket()
6      throws SocketException { super(); }
7
8    // Overriding the base-level methods.
9    public void send(java.net.DatagramPacket p0)
10     throws IOException {
11     if(metaObject == null)
12     { super.send(p0); return; }
13     ...
14     Class[] paramType = new Class[1];
15     paramType[0] = java.net.DatagramPacket.class;
16     Method method = WrapperLevel_MulticastSocket.
17       class.getDeclaredMethod("send", paramType);
18
19     Object[] tempArgs = new Object[1];
20     tempArgs[0] = p0;
21     ChangeableBoolean isReplyReady =
22       new ChangeableBoolean(false);
23
24     try {
25       metaObject.invokeMetaMethod
26         (method, tempArgs, ... );
27     } catch (java.io.IOException e) { throw e; }
28     catch (MetaMethodIsNotAvailable e) {}
29   }
```

Fig. 16.   Excerpted generated wrapper code.

to call one or more of the base-level methods. To support such cases, which we suspect might be very common, the wrapper-level class provides access to the base-level methods through the special wrapper-level methods whose names match the base-level method names, but with an "Orig_" prefix.

Generated Metalevel Class. Figure 17 shows excerpted code for MetaLevel_MulticastSocket, the generated metalevel class for Multicast-Socket. This class keeps an instance variable, delegates, which is of type Vector and refers to all the delegate objects associated with a metaobject that implements one or more of the base-level methods. To support dynamic adaptation of the static methods, a metalevel class provides the staticDelegates instance variable and its corresponding insertion and removal methods (not shown). *Delegate* classes introduce new code to applications

```
 1   public class MetaLevel_MulticastSocket
 2     extends UnicastRemoteObject
 3     implements MetaLevel_Interface,DelegateManagement{
 4
 5     private Vector delegates = new Vector();
 6     public synchronized void insertDelegate
 7       (int i, String delegateClassName)
 8       throws RemoteException { ... }
 9     public synchronized void removeDelegate(int i)
10      throws RemoteException { ... }
11
12     public synchronized Object invokeMetaMethod
13       (Method method, Object[] args,
14       ChangeableBoolean isReplyReady) throws Throwable{
15       // Finding a delegate that implements this method
16       ...
17       if(!delegateFound) // No meta-level method available
18         throw new MetaMethodIsNotAvailable();
19       else
20         return newMethod.invoke(delegates.get(i-1),
21           tempArgs);
22   }
```

Fig. 17. Excerpted generated metaobject code.

at run time by overriding a collection of base-level methods selected from one or more of the *adaptable* base-level classes. An adaptable base-level class has corresponding wrapper- and metalevel classes, generated by TRAP/J at compile time. Metaobjects can be programmed dynamically by inserting or removing delegate objects at run time. To enable a user to change the behavior of a metaobject dynamically, the metalevel class implements the DelegateManagement interface, which in turn extends the Java RMI Remote interface (lines 5–10). A composer can remotely "program" a metaobject through Java RMI. The insertDelegate and removeDelegate methods are developed for this purpose.

The metaobject protocol developed for metalevel classes defines only one method, invokeMetaMethod, which first checks if any delegate is associated with this metaobject (lines 12–22). If not, then a MetaMethod-IsNotAvailable exception is thrown, which eventually causes the wrapper method to call the base-level method as described before. Alternatively, if one or more delegates is available, then the first delegate that overrides the method is selected, a new method on the delegate is created using Java reflection, and the method is invoked.

Adapting to Loss Rate. To evaluate the TRAP/J-enhanced audio application, we conducted two sets of experiments similar to those in the previous section. The configuration used in these sets of experiments is illustrated in Fig. 12.

In the first sets of experiments, a user holding a receiving iPAQ handheld computer is walking within the wireless cell, receiving and playing a live audio stream. Figure 18 shows a sample of the results. For the first 120 seconds, the program has no FEC capability. At 120 seconds, the user walks away from the sender and enters an area with loss rate around 30%. The adaptable application detects the high loss rate and inserts a (4, 2) FEC filter, which greatly reduces the packet loss rate as observed by the application, and improves the quality of the audio as heard by the user. At 240 seconds, the user approaches the sender, where the network loss rate is again low. The adaptable application detects the improved transmission and removes the FEC filters, avoiding the waste of bandwidth with redundant packets. Again at 360 seconds, the user walks away from the sender, resulting in the insertion of FEC filters. This experiment demonstrates the utility of TRAP/J to transparently and automatically enhance an existing application with new adaptive behavior.

Balancing QoS and Energy Consumption. In the second set of experiments, we used two MetaSocket filters, SendNetLossDetector and RecvNetLossDetector, which cooperate to monitor the raw loss rate of the

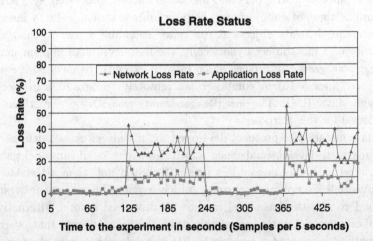

Fig. 18.   The effect of using FEC filters to adapt ASA to high loss rates on a wireless network.

wireless channel. Similarly, the SendAppLossDetector and RecvAppLossDetector filters are used to monitor the packet loss rate as observed by the application, which may be lower than the raw packet loss-rate due to the use of FEC. At present, a simple state machine is used by a decision maker (DM) component to govern changes in filter configuration. For example, if the loss rate observed by the application rises above a specified threshold, then the DM decides to insert an FEC filter in the pipeline. In case an FEC filter is already present in the pipeline, DM decides to modify the (n, k) parameters of the FEC filter to increase improve QoS. On the other hand, if the raw packet loss rate on the channel drops below a lower threshold, then the level of redundancy is decreased by modifying the parameters of the FEC filter, or in case the FEC filter is not required anymore, DM removes the FEC filter entirely.

Figure 19 shows a trace of an experiment using the ASA described earlier, running in *ad hoc* mode. A stationary user speaks into a laptop microphone, while another user listens on an iPAQ as he changes his location in the wireless cell over a period of time. In this particular test, the iPAQ user remains in a low packet loss area for approximately 30 minutes, moves to a high packet loss area for another 40 minutes, moves back to the low packet loss location for another 30 minutes, and then re-enters the high packet loss location. The user remains there until the iPAQ's external battery drains and the network is disconnected.

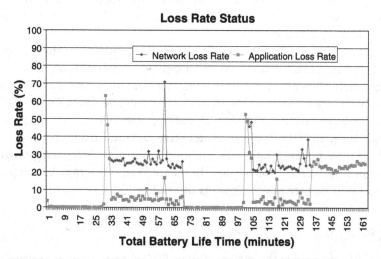

Fig. 19. MetaSocket packet loss behavior with dynamic FEC filter insertion and removal.

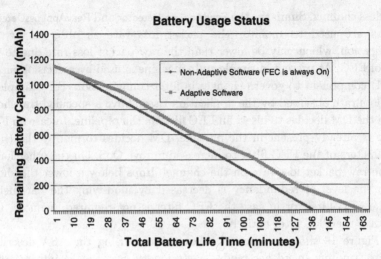

Fig. 20.   Trace of energy consumption during experiment using a software measurement technique.

In this experiment, the upper threshold for the RecvAppLossDetector to generate an UnAcceptableLossRateEvent is 20%, and the lower threshold for the RecvNetLossDetector to generate an AcceptableLossRateEvent is 5%. As shown in Fig. 19, the FEC $(4, 2)$ code is effective in reducing the packet loss rate as observed by the application. Figure 20 plots the remaining battery capacity as measured during the above experiment and that for a non-adaptive trace. The adaptive version extends the battery lifetime by approximately 27 minutes.

In summary, TRAP enables production of adaptable program families from existing programs developed in class-based, object-oriented programming languages. Using the wrapper- and metalevel classes as hooks instrumented inside the application code at compile time, TRAP enables separate development and deployment of adaptive code in existing programs at run time. In TRAP, pieces of adaptive code are realized as delegates that can be inserted into and removed from an adapt-ready program dynamically, thereby converting the adapt-ready program to adaptable programs in its corresponding program subfamily.

6. Discussion

Figure 21 summarizes the current status of transparent shaping realizations. We have implemented and tested ACT/J and TRAP/J, as described above.

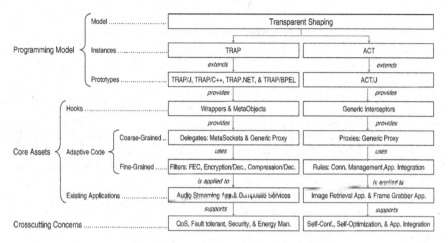

Fig. 21. Transparent shaping status.

We have also developed several core assets for supporting transparent shaping, including examples of hooks, adaptive code, and existing applications. The hooks in TRAP/J are pairs of wrappers and metaclasses, which are generated by TRAP/J generators automatically. In ACT/J, there is only one hook, the generic portable interceptor, which can be reused in any CORBA program. Adaptive code in TRAP/J is realized by delegates. A reusable delegate using MetaSockets and filters is provided. A generic proxy was developed for ACT/J that can be used in any existing CORBA application. The generic proxy can receive any CORBA request and can adapt it using adaptive code realized by rules.

We are currently addressing several other aspects of transparent shaping. To support existing programs developed in C++, .NET, and BPEL, members of our group have already implemented TRAP/C++[31] using compile-time metaobject protocols supported by Open C++,[32] TRAP.NET[33] using a combination of reflective capabilities in C# and Microsoft common intermediate language (CIL), and TRAP/BPEL[34] by wrapping the invocations and forwarding them to a local generic proxy developed in Java, respectively. To support CORBA programs developed using C++ ORBs, we plan to develop ACT/C++. We are also investigating techniques to support the insertion of hooks for adaptation into the operating system kernel,[19] the third case mentioned earlier.

Transparent shaping complements other work in adaptive software, particularly adaptive middleware. Figure 22 depicts this relationship,

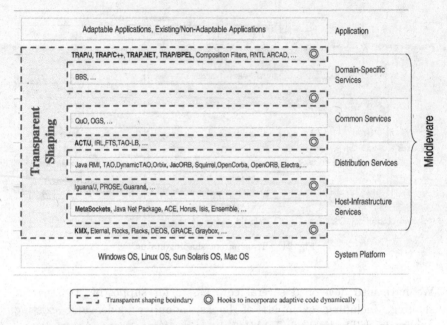

Fig. 22.   Relationship of transparent shaping to other contributions.

according to Schmidt's four-layer middleware taxonomy.[35] Please note
that the frameworks mentioned inside the transparent shaping boundary
can be incorporated into existing applications transparently, while the
ones outside this boundary require explicit calls from the application
source code. As in our work with TRAP/J and MetaSockets, transparent
shaping can enable existing non-adaptive applications to take advantage
of adaptive host-infrastructure middleware services such as MetaSockets.
Also, using our ACT/J framework, transparent shaping can enable existing
CORBA applications to take advantage of adaptive common middleware
services such as QuO. In addition, we note that many adaptive frameworks
developed by other groups can be used to support transparent shaping.
Examples include Composition Filters,[36] RNTL ARCAD,[10] Interoperable
Replication Logic,[37] FTS,[38] TAO Load Balancing,[39] Iguana/J,[40] Prose,[41]
Guaranà,[42] Eternal,[43] and Rocks/Racks.[44] Previously, we provided a
summary of these and several other techniques.[1]

Finally, we note that transparent shaping has potential impact
beyond supporting adaptation in individual programs, for example, to
support application integration.[45] To integrate two existing heterogeneous

applications, possibly developed in different programming languages and targeted to run on different platforms, one needs to convert data and commands between the two applications on an ongoing basis. Transparent shaping offers a solution to this problem, without the need to modify application source code directly. In preliminary studies,[45,46] we have proposed several alternative architectures and showed how transparent shaping can support interoperability, via Web services, for Java RMI, CORBA, and .NET applications. As a proof of concept, we have conducted a case study that demonstrates the use of transparent shaping in the integration of an image retrieval application developed in CORBA with a frame grabber application developed in .NET.

7. Conclusions and Future Work

Transparent shaping supports reuse of existing programs in new, dynamic environments even though the specific characteristics of such new environments were not necessarily anticipated during the original design of the programs. In particular, many existing programs, not designed to be adaptable, are being ported to dynamic wireless environments, or hardened in other ways to support pervasive and autonomic computing. We have described an approach to transparent shaping based on the concept of program families and demonstrated how automated methods can be used to transform a program into another member of the same family. Our approach integrates four key technologies: aspect-oriented programming, behavioral reflection, component-based programming, and adaptive middleware. We highlighted two different realizations of transparent shaping, ACT and TRAP, and showed how they realize the general adaptive programming model. In addition to our work on other realizations of transparent shaping, as well as application integration, we are also addressing several other aspects of transparent shaping: coordination of adaptive behavior across system layers and among different systems, formal techniques to ensure that adaptations leave the system in a consistent state,[47] preventing adaptation mechanisms from being exploited by would-be attackers, and constructing "product lines" of adaptable software.

Acknowledgements

We express our gratitude to the faculty and students in the Software Engineering and Network Systems Laboratory at Michigan State University for their feedback and their insightful discussions on this work. This work was supported in part by the US Department of the Navy, Office of Naval Research under Grant No. N00014-01-1-0744, and in part by National Science Foundation Grants CCR-9912407, EIA-0000433, EIA-0130724, ITR-0313142, and CCR-9901017.

References

1. McKinley, P.K., Sadjadi, M., Kasten, E.P., Cheng, B.H.C.: Composing adaptive software, *IEEE Computer*, (July, 2004), pp. 56–64.

2. Weiser, M.: Ubiquitous computing, *IEEE Computer*. 26 (10), (October, 1993), pp. 71–72. ISSN 0018–9162.

3. Kephart, J.O., Chess, D.M.: The vision of autonomic computing, *IEEE Computer*, 36(1), 2003, pp. 41–50, ISSN 0018–9162.

4. Parnas, D.L.: On the design and development of program families, *IEEE Transactions on Software Engineering* (March. 1976).

5. Tarr, P., Ossher, H.: Eds. *Workshop on Advanced Separation of Concerns in Software Engineering at ICSE 2001 (W17)* (May, 2001).

6. Kiczales, G., Lamping, J., Mendhekar, A., Maeda, C., Videira Lopes, C., Loingtier, J.M., Irwin, J.: Aspect-oriented programming. In *Proceedings of the European Conference on Object-Oriented Programming (ECOOP)*. Springer-Verlag LNCS 1241 (June, 1997).

7. *Communications of the ACM, Special Issue on Aspect-Oriented Programming*, (October, 2001), Vol. 44.

8. Maes, P.: Concepts and experiments in computational reflection. In *Proceedings of the ACM Conference on Object-Oriented Languages (OOPSLA)*, ACM Press (December, 1987), pp. 147–155, ISBN 0-89791-247-0.

9. Kiczales, G., des Rivières, J., Bobrow, D.G.: *The Art of Metaobject Protocols*. (MIT Press, 1991).

10. David, P.C., Ledoux, T., Bouraqadi-Saadani, N.M.N.: Two-step weaving with reflection using AspectJ. In *OOPSLA 2001 Workshop on Advanced Separation of Concerns in Object-Oriented Systems*, Tampa (October, 2001).

11. Szyperski, C.: *Component Software: Beyond Object-Oriented Programming*, (Addison-Wesley, 1999).

12. Bakken, D.E.: *Middleware* (Kluwer Academic Press, 2001).

13. *Proceedings of the Middleware'2000 Workshop on Reflective Middleware (RM2000)*, New York (April, 2000).

14. Yang, Z., Cheng, B.H.C., Stirewalt, R.E.K., Sowell, J., Sadjadi, S.M., McKinley, P.K.: An aspect-oriented approach to dynamic adaptation. In *Proceedings of the ACM SIGSOFT Workshop On Self-healing Software (WOSS'02)* (November, 2002).

15. *The Common Object Request Broker: Architecture and Specification Version 3.0*. Object Management Group, Framingham, Massachusett (July, 2003).

16. Cohen, G.A., Chase, J.S., Kaminsky, D.: Automatic program transformation with JOIE. In *1998 Usenix Technical Conference* (June, 1998).

17. Adve, S., Harris, A., Hughes, C., Jones, D., Kravets, R., Nahrstedt, K., Sachs, D., Sasanka, R., Srinivasan, J., and Yuan, W.: The Illinois GRACE project: Global resource adaptation through cooperation, 2002.
18. Distributed extensible open systems (the DEOS project), (2004). Georgia Institute of Technology — College of Computing.
19. Samimi, F., McKinley, P.K., Sadjadi, S.M., Ge, P.: Kernel-middleware cooperation in support of adaptive mobile computing. In *the Second International Workshop on Middleware for Pervasive and Ad-Hoc Computing*.
20. Sadjadi, S.M., McKinley, P.K.: ACT: An adaptive CORBA template to support unanticipated adaptation. In *Proceedings of the 24th IEEE International Conference on Distributed Computing Systems (ICDCS'04)*, Tokyo, Japan (March, 2004).
21. Sadjadi, S.M., McKinley, P.K.: Transparent self-optimization in existing CORBA applications. In *Proc. of the International Conference on Autonomic Computing (ICAC-04)*, (May, 2004), pp. 88–95, New York, NY.
22. *ORBacus for C++ and Java version 4.1.0*. IONA Technologies Inc., (2001).
23. Brose, G., Noffke, N.: JacORB 1.4 documentation. Technical report, Freie Universitt Berlin and Xtradyne Technologies AG (August, 2002).
24. Schmidt, D.C., Levine, D.L., Mungee, S.: The design of the TAO real-time object request broker, *Computer Communications*. 21(4) (April, 1998), pp. 294–324.
25. Sadjadi, S.M., McKinley, P.K.: Supporting transparent and generic adaptation in pervasive computing environments. Technical Report MSU-CSE-03-32, Department of Computer Science, Michigan State University, East Lansing, Michigan (November, 2003).
26. Zinky, J., Loyall, J., Shapiro, R.: Runtime performance modeling and measurement of adaptive distributed object applications. In *Proceedings of the International Symposium on Distributed Object and Applications (DOA 2002)*, Irvine, California (October, 2002).
27. Sadjadi, S.M., McKinley, P.K., Cheng, B.H.C., Stirewalt, R.K.: TRAP/J: Transparent generation of adaptable Java programs. In *Proceedings of the International Symposium on Distributed Objects and Applications (DOA'04)*, Agia Napa, Cyprus (October, 2004).
28. Kiczales, G., Hilsdale, E., Hugunin, J., Kersten, M., Palm, J., Griswold, W.G.: An overview of AspectJ, *Lecture Notes in Computer Science*, 2072, 2001, pp. 327–355.
29. Piveta, E.K., and Zancanella, L.C.: Aspect weaving strategies, *Journal of Universal Computer Science*, 9(8), 2003, pp. 970–983.
30. Rizzo, L., and Vicisano, L., RMDP: An FEC-based reliable multicast protocol for wireless environments, *ACM Mobile Computer and Communication Review*. 2(2) (April, 1998).
31. Fleming, S.D., Cheng, B.H.C., Stirewalt, R.K., and McKinley, P.K.: An approach to implementing dynamic adaptation in C++. In *Proceedings of the first Workshop on the Design and Evolution of Autonomic Application Software 2005 (DEAS'05), in conjunction with ICSE 2005*, St. Louis, Missouri (May, 2005).

32. Chiba, S., Masuda, T.: Designing an extensible distributed language with a meta-level architecture, *Lecture Notes in Computer Science*, 707, 1993.

33. Sadjadi, S.M., Trigoso, F.: TRAP.NET: A realization of transparent shaping in .net. In *Proceedings of The Nineteenth International Conference on Software Engineering and Knowledge Engineering (SEKE'2007)*, pp. 19–24, Boston, USA (July, 2007).

34. Ezenwoye, O., Sadjadi, S.M.: TRAP/BPEL: A framework for dynamic adaptation of composite services. In *Proceedings of the International Conference on Web Information Systems and Technologies (WEBIST 2007)*, Barcelona, Spain (March, 2007).

35. Schmidt, D.C.: Middleware for real-time and embedded systems, *Communications of the ACM*, 45(6) (June, 2002).

36. Bergmans, L., Aksit, M.: Composing crosscutting concerns using composition filters, *Communications of ACM.* (10) (October, 2001), pp. 51–57.

37. Baldoni, R., Marchetti, C., Termini, A.: Active software replication through a three-tier approach. In *Proceedings of the 22th IEEE International Symposium on Reliable Distributed Systems (SRDS02)*, pp. 109–118, Osaka, Japan (October, 2002).

38. Hadad, E.: *Architectures for Fault-Tolerant Object-Oriented Middleware Services*. PhD thesis, Computer Science Department, The Technion — Israel Institute of Technology, 2001.

39. Othman, O.: The design, optimization, and performance of an adaptive middleware load balancing service. Master's thesis, University of California, Irvine, 2002.

40. Redmond, B., Cahill, V.: Supporting unanticipated dynamic adaptation of application behaviour. In *Proceedings of the 16th European Conference on Object-Oriented Programming* (June, 2002).

41. Popovici, A., Gross, T., Alonso, G.: Dynamic homogeneous AOP with PROSE. Technical Report, Department of Computer Science, Federal Institute of Technology, Zurich, 2001.

42. Oliva, A., Buzato, L.E.: The implementation of Guaraná on Java. Technical Report IC-98-32, Universidade Estadual de Campinas (September, 1998).

43. Moser, L., Melliar-Smith, P., Narasimhan, P., Tewksbury, L., Kalogeraki, V.: The Eternal system: An architecture for enterprise applications. In *Proceedings of the Third International Enterprise Distributed Object Computing Conference (EDOC'99)* (July, 1999).

44. Zandy, V.C., Miller, B.P.: Reliable network connections. In *Proceedings of the Eighth Annual International Conference on Mobile Computing and Networking* (September, 2002), pp. 95–106.

45. Sadjadi, S.M.: *Transparent Shaping for Existing Software to Support Pervasive and Autonomic Computing*. Ph.D. thesis, Department of Computer Science, Michigan State University, East Lansing, United States (August, 2004).

46. Sadjadi, S.M., McKinley, P.K.: Using transparent shaping and web services to support self-management of composite systems. In *Proceedings of the*

International Conference on Autonomic Computing (ICAC'05), Seattle, Washington (June, 2005).

47. Zhang, J., Yang, Z., Cheng, B.H.C., McKinley, P.K.: Adding safeness to dynamic adaptation techniques. In *Proceedings of the ICSE 2004 Workshop on Architecting Dependable Systems*, Edinburgh, Scotland (May, 2004).

Chapter 5

RULE EXTRACTION TO UNDERSTAND CHANGES
IN AN ADAPTIVE SYSTEM

MARJORIE A. DARRAH

Department of Mathematics, West Virginia University
Morgantown, WV26505, USA
mdarrah@math.wvu.edu

BRIAN J. TAYLOR

Department of Computer Science, University of Massachusetts
Amherst, MA 01003, USA
btaylor@cs.umass.edu

Rule extraction as a specific technique can support the rigorous development and assurance needed before using adaptive systems in mission- and safety-critical applications. The ultimate goal and significant innovation of this research discussed in this chapter is to demonstrate that neural network rule extraction technology could be transferred into a practical software tool for neural network verification and validation and other purposes, such as monitoring the state of a neural network. Such a tool would accept as input a formal specification of the trained neural network and use neural network rule extraction algorithms to translate the neural network into an equivalent set of rules. These rule-based systems, which represent the neural network's knowledge, have a more visible and potentially human-readable decision logic that supports a robust set of verification techniques. Rule extraction technology in the form of a usable tool will dramatically increase the ability to verify and validate high assurance neural network systems.

1. Neural Network Rule Extraction

Neural networks are members of a class of software well suited for domains of non-linearity and high complexity that are ill defined, unknown, or just too complex for standard programming practices. Verifying correct operation of neural networks within projects such as autonomous mission control agents, vehicle health monitoring systems, adaptive flight controllers, or nuclear engineering applications requires a rigorous approach.

Testing the neural network with data like that used in training is one of a few methods used to verify that a network has adequately learned the input domain. In non-critical applications, such traditional testing techniques prove adequate for the acceptance of a neural network system that has been trained using input conditions that do not vary from operational conditions. However, in more complex, safety- and mission-critical systems, the standard neural network training-testing approach alone will not provide a reliable method for their certification.

The verification and validation challenge is further compounded by adaptive neural network systems that modify themselves, or "learn", during operation. Traditional software assurance methods fail to account for systems that change after deployment.

This chapter outlines the investigation of a technique known as neural network rule extraction to determine its usefulness toward the verification and validation of neural networks in safety-critical applications. Rule extraction, in general, is the process of developing English-like syntax that describes the internal knowledge of the neural network. Rule extraction can be used as a method for verifying the neural network knowledge acquisition processes. It is a technique that translates the decision process of a trained neural network into an equivalent decision process represented as a set of rules.

The techniques of rule extraction have been used to model the knowledge that the neural network has gained while training or adapting. The rules extracted are generally represented by a set of if-then statements that may be examined by a human. If the neural network is fixed after training then the rules should, with some confidence level, model the way the neural network will handle other data that is processed. If the neural network is an online adaptive neural network, then rule extraction can be done for one point of time to establish what the system looks like at that instance. Repeated application of rule extraction could yield an understanding of the progression of the network during adaptation.

1.1. *Background on Rule Extraction*

Much research is ongoing in the area of rule extraction and a literature survey uncovered many useful algorithms and techniques. The techniques developed thus far are neural network specific. Nearly 200 recent artifacts related to neural network rule extraction were identified. These artifacts include IEEE and other major journal articles, conference proceedings, academic dissertations and theses, and technical reports. The purpose of the literature survey was to identify rule extraction algorithms and techniques that could be included in a general rule extraction tool. From the survey of research, it is clear that this technology has progressed to the point that it may now be realized in a comprehensive tool for commercial use.

Because of the abundance of approaches to rule extraction, the techniques were categorized based on three criteria:

1. Translucency of rules
   a. Pedagogical (blackbox)
   b. Decompositional (whitebox)
   c. Eclectic (mixed approach)
2. Neural Network portability
3. Type of rules generated
   a. Fuzzy
   b. Boolean

A summary of techniques was developed that includes the rule translucency, the neural network portability, and the rule format. A table of this information is provided in Appendix A. Many of the techniques are network specific. The main types of neural networks in use include feed forward, multilayer perceptrons, radial basis function, and self-organizing maps.

The focus of this research was to determine how decompositional (specific to neural network) and pedagogical (applicable to many neural networks) techniques can be applied to widely used neural network types. A decompositional technique is a "whitebox" approach that extracts rules using a neuron-by-neuron series of steps. The advantage of the decompositional approach is that it offers the prospect of generating a complete set of rules for the neural network. Future versions of a general tool for rule extraction should include decompositional techniques related to the most commonly used neural networks. Two of the most promising techniques that fit this category are the techniques developed for self-organizing maps (discussed later) and the techniques for feedforward neural networks presented by Setiono [2002]. Both of these techniques apply to neural networks used for function approximation and extract very accurate rule sets.

In the future, the rule extraction tool should also include at least one pedagogical approach that applies to all neural network types. A pedagogical technique is a "blackbox" approach. In this type of approach the neural network is trained on the data, and then the pedagogical algorithm considers the training data and the associated neural network output to develop a rule structure based only on the data without consideration of the neural network type. Although the pedagogical approaches try to mimic the performance of the neural network, they are somewhat less likely to accurately capture all of the valid rules describing a network's knowledge. Their usefulness lies in the fact that these algorithms apply to most neural network types. The techniques that could be used for this approach include a genetic algorithm called GREX developed by Johansson [2004] or the TREPAN algorithm developed by Craven [1994].

2. Rule Extraction for System Verification and Validation

In studies of extraction processes, an inverse relationship exists between the degree of determinism and the readability of the rules produced. To be most useful for verification and validation, both readability for domain expert review and determinism for automated model analysis or execution, are desirable. Thus the extraction process must accommodate these needs either through parameter-based means or via separate algorithms.

Rule bases generated for domain expert review should sacrifice determinism for readability where necessary. It is desirable that the degree of sacrifice be tunable on either an input or output basis so that important interfaces may be more thoroughly covered. It is preferred that the tuning approach provides a target level of coverage or determinism for the algorithm to seek. If this type of tuning is not possible, a reasonable alternative would be to report the percentage of coverage or determinism achieved so that the operator could choose to reanalyze at other levels if necessary.

Rule bases generated for automated model analysis or for execution should sacrifice readability for determinism. It is not necessary that the degree of this sacrifice be tunable. The format of the deterministic rule bases should be suitable for providing the rule portion of the input to commercial-off-the-shelf (COTS) automated model analysis tools. A possible target is the SSL format utilized by the SCRTool, a toolset created by the Naval Research Laboratories for formal analysis of specification models. This format is widely used by other automated model analysis tools.

Examination of rule extraction from a verification and validation perspective led to the following summary of observations for the rule extraction process and requirements for two rule types, readable and deterministic.

Observations for Rule Extraction Process

- Rule-base consumers include domain expert review, automated model analysis and rule base execution.
- Rule-base consumers can be classified by need for readable rule bases vs. deterministic rule bases.
- An inverse relationship may exist between the degree of determinism and the readability of the rule bases.
- Either separate or tunable rule extraction processes must be developed to target these classifications.

Readable Rule-Base Requirements

- Should be understandable by domain experts who are not rule base or neural network experts.
- When a tradeoff is made between determinism and readability, the degree of tradeoff should be adjustable.
- It is desirable to be able to adjust the tradeoff for targeted I/O domains as well as the whole neural network.

Deterministic Rule-Base Requirements

- Deterministic rule bases shall be fully deterministic regardless of whether that determinism fully agrees with the neural network If the rule bases do not fully agree with the neural network, it is desirable both to control the degree of the agreement and to report characterization of the agreement.
- The rule base shall adhere to the format of COTS tools (such as SCRTool) that consume models that are either rule bases or supersets of rule bases that will be used for analysis.

2.1. *An Example of Rule Extraction for the Dynamic Cell Structure Neural Network Used in a System*

Previous work done at the Institute for Scientific Research, Inc. (ISR) developed a rule extraction technique for the dynamic cell structure (DCS) neural network [Darrah 2004]. The DCS is a type of self-organizing map (SOM) and is a component of the Intelligent Flight Control System (IFCS) developed by NASA Dryden Flight Research Center, NASA Ames Research

Center and others.[1] The DCS algorithm was originally developed by Bruske and Sommer [1994] and is a derivative of work by Fritzke [1994] combined with competitive Hebbian learning by Martinez [1993]. These neural networks were designed as topology representing networks whose roles are to learn the topology of an input space.

The DCS neural network partitions the input space into Voronoi regions.[2] The neurons within the neural network represent the reference vector (centroid) for each of the Voronoi regions. The connections between the neurons, c_{ij}, are then part of the Delaunay triangulation[3] connecting neighboring Voronoi regions through their reference vectors.

Given an input, v, the best matching unit (BMU) is the neuron whose weight, w, is closest to v. Along with the BMU, the neighbors of the BMU are found through the Delaunay triangulation. During adaptation, adjustments are made to the BMU and neurons within the BMU neighborhood based on the input.

The DCS algorithm consists of two learning rules, Hebbian and Kohonen. These two learning rules allow the DCS neural network to change its structure to adapt to inputs. The ability to adjust neuron positions and add new neurons into the network gives the DCS neural network the potential to evolve into many different configurations. Figure 1 shows the first basic algorithm developed for extracting human readable rules from the DCS.

2.1.1. *Refining the Algorithm*

The original rule extraction process and the rule type were closely examined for usability requirements. The objectives for refining the DCS rule extraction algorithm were to:

- Increase accuracy.
- Maintain human understandability.
- Provide deterministic rule base that could be input to a commercial-off-the-shelf (COTS) tool or implemented as a rule-based system.

---

[1]The Intelligent Flight Control System is being developed by the NASA DFRC, NASA ARC, Boeing Phantom Works, the Institute for Scientific Research, Inc., and West Virginia University.

[2]Given a set of n points in the plane, a Voronoi partition is a collection of n convex polygons such that each polygon contains exactly one point and every point in a given polygon is closer to its central point than to any other.

[3]The Delaunay triangulation is a dual graph of a Voronoi diagram that connects the centers of the Voronoi regions to their neighbors to form a triangulation of the plane (if no two points are cocircular).

Input:
      Weights from a trained DCS (centers of Voronoi region)
      Best matching unit for each input
Output:
      One rule for each cell of the DCS
Procedure:
      Apply input stimulus to DCS from training data
      Record BMU for each input
      Collect all inputs with common BMU to form cell
      For each weight (w_i)
            For each independent variable
                  $x_{lower} = \min\{x \mid x \text{ has BMU} = w_i\}$
                  $x_{upper} = \max\{x \mid x \text{ has BMU} = w_i\}$
      Build rule by:
            Independent variable in $[x_{lower}, x_{upper}]$
            Join antecedent statements with AND
            Dependent variable = category
                  OR
            Dependent variable in $[y_{lower}, y_{upper}]$
            Join conclusion statements with AND
      Write Rule

Fig. 1.   Human understandable rule extraction algorithms for DCS.

It was determined that two separate algorithms would be necessary to achieve both human understandability and determinism. A refinement of the original algorithm was made to improve the human readable rules and a new algorithm was developed to generate deterministic rules. Both algorithms utilize the structure of the DCS knowledge by considering the Voronoi regions that partition the input space.

As explained previously, the DCS partitions the input space into Voronoi regions. These regions are convex polygons in two dimensions and convex n-dimensional polyhedra in n dimensions. The original rule extraction algorithm did not capture the entire polygon or polyhedron region with the rules. The original algorithm used a "box" to represent that region and the rules represent the box. (See Fig. 2.)

A new rule extraction algorithm was developed to completely capture the polygonal regions of the input space that represent the structure of the trained DCS. The previous rule format was non-deterministic and although understandable, could not be used as input to COTS tools or implemented.

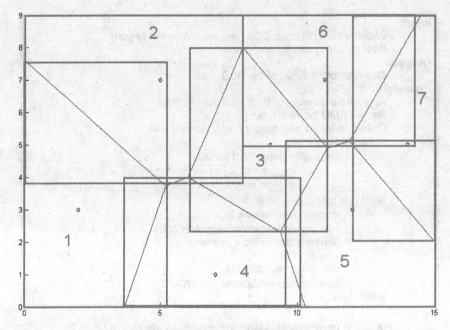

Fig. 2.   Original rule coverage.

The new rule format is

    IF input ∈ region 1 (input satisfies a set of constraints)

    THEN output = multivariable linear expression

Below is an example of the new deterministic rules for a two-dimensional data set.

```
IF (6*x + 0*y >= 48) AND (2*x + 2*y >= 32)
    AND (-1*x + 4*y >= 8.5) AND (-3*x + 2*y >= -25.5)
    AND (4*x - 2*y >= 16) AND (-3*x + 2*y >= -23.5)
    AND (-5*x + 0*y >= -57.5)
THEN z = 0.75*x + 0.75*y - 7.5
ENDIF
```

These rules are specifically designed for the DCS structure implemented by the IFCS project. The first part of the rule (after the IF) gives a set of constraints that defines a region of the input space based on a possible BMU and second best matching unit (SEC) pair. The subsequent parts

---

Inputs:
       , P = Set of all weights (centroids of the Voronoi regions)
        A = Adjacency Matrix
Output:
        R = Set of rules that describe a partition of the input space with
        associated outputs
Procedure:
        Use P to define a Voronoi diagram that partitions the input space.
        Use A to determine neighboring regions in the Voronoi diagram to find
        BMU and SEC pairs.
    For each p ∈ P (centroid of Voronoi region and BMU)
        Calculate Voronoi region boundaries.
        For each q ∈ P – {p} such that p is a neighbor of q (centroid of
        neighboring regionand SEC)
        Determine boundaries that divide the region with centroid w into
        subregions.
        Determine antecedent based on boundaries defined by p and q.
           Determine consequent equation based on DCS output
           determined by p and q.
           Write rule.

Fig. 3.   Deterministic rule extraction algorithm.

(after the THEN) yield the DCS output based on this region. For the IFCS DCS implementation, the output is determined based on the BMU and SEC pair.

The algorithm for the deterministic rule extraction technique for the DCS is shown in Fig. 3. This algorithm was developed because no such technique existed for self-organizing maps. However, a similar rule extraction technique does exist for feedforward neural network. The techniques used to create the rule extraction process for the DCS align with the techniques developed by Setiono [2002] and outlined in his paper "Extraction of rules from artificial neural networks for regression."

To test the accuracy, deterministic rules were generated for three different data sets. The rule output and the neural network output had 100% agreement. The rules are constructed to completely cover the input space and to use the DCS recall function as the output based on the region, therefore these rule have complete agreement with the neural network. Figure 4 below shows a two-dimensional example of how the rules partition the data based on BMU and SEC. The solid lines in the figure indicate the original Voronoi regions that divide the plane based on the BMU. The dotted lines show how the original regions are subdivided to account for the SEC. The lines that define the subregions form the rule antecedent.

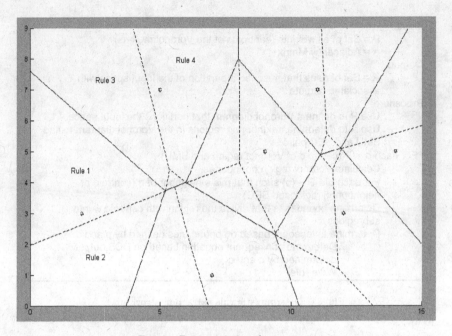

Fig. 4.   Deterministic rule coverage.

Note that the entire input space is covered, with each of the subregions representing one rule.

Once the deterministic rule algorithm was developed it could then be used to help to improve the accuracy of the human readable rules. The human readable rules were generated by the original algorithm, which defined the boundaries of the rules by using the training data and its associated output. Since the deterministic rules agreed with the DCS 100%, more input/output data could be generated and fed to the original algorithm. The original algorithm can then generate human understandable rules that more fully cover the input space. The rules are not completely accurate and overlap occurs in the rules, but they are improved and still readable.

3. Applying Rule Extraction in a Tool for Verification and Validation

In order to make rule extraction usable for the verification and validation practitioner, there must be a way to facilitate its use with any neural network in a way that does not require a high degree of sophisticated research for its application. This section outlines the steps that can be done to facilitate this process.

3.1. *Describing a Neural Network with Metadata Expressions*

Several metadata expressions for neural networks were identified. These include Neural Network Markup Language (NNML), NeuroML, and Matlab Simulink.

Neural Network Markup Language (NNML) is an XML-based language for the description of neural network models [Robtsov 2004]. The goal of NNML is to enable the description of a neural network model completely, including data dictionary, pre- and post-processing, details of structure and parameters, as well as other auxiliary information. NNML facilitates the exchange of neural network models as well as documenting, storing, and manipulating them independently from the system that produced them.

NeuroML is another XML-based markup language for describing models of neural systems [Goddard 2001]. It is intended as an interchange format for software tools in use — to allow different programs to cooperate, build, simulate, test, and publish models. The NeuroML specification describes not only the model specifications in NeuroML, but also the run time interfaces that simulators can implement to expose instantiated model state. There is a NeuroML development kit including software for reading and writing NeuroML models. The specification also describes how to extend the vocabulary of NeuroML to accommodate new modeling techniques.

Matlab/Simulink offers a graphical programming language that can describe a neural network design. It provides the neural network developer the ability to use neural network construction blocks, inter-block connections and pre-defined mathematical functions (including prominent neural network learning algorithms, basic statistical functions, and common computations). While the Matlab/Simulink language would require the usage of the Matlab application, it appears to be commonly used among the research community and would provide a ready-to-use development environment.

These three metadata description methods were studied to determine which one could best be incorporated in a rule extraction process and prototype. NeuroML appeared more suited for description of biological neural networks. Matlab description requires that developers only use the Matlab environment. Therefore, NNML was chosen because it dealt with artificial neural networks and because it was platform and language independent.

3.2. *Building a Tool for Rule Extraction*

An overview of the general neural network rule extraction tool process is depicted in Fig. 5 and described below.

Input File. The file will be accepted in NNML format.

Parse NNML File. The NNML data file will be parsed and the relevant data needed by the individual rule extraction algorithms will be retained.

Extract Rules. The parsed NNML file will be input to the rule extraction algorithm. Each algorithm is specific to the neural network type and will require various inputs. There may be more than one algorithm for a specific neural network type.

Display Rule Expression. The rule expression will be presented to the user in a human readable format and deterministic format.

Analyze Rules. The rules, in a tool understandable format, will be used as input to an outside tool such as SCRTool to check for rule consistency. Test vectors will be generated for use in testing the neural network.

Generate Report. A report on the domain coverage and rule consistency will be generated.

Display Analysis Report. The analysis report will be displayed to the user in a graphical format.

Execute Rules. The rules will be input and converted to a form for execution. Test vectors will be input and used for testing the rules.

Execute Neural Network. Test vectors will be input and used for testing the NN.

Compare Results. Output from the rules and the neural network will be compared for agreement and possibly accuracy against a known set.

Generate Report. A report on rule vs. neural network agreement and possibly accuracy will be generated.

Display Comparison Report. The comparison report will be displayed.

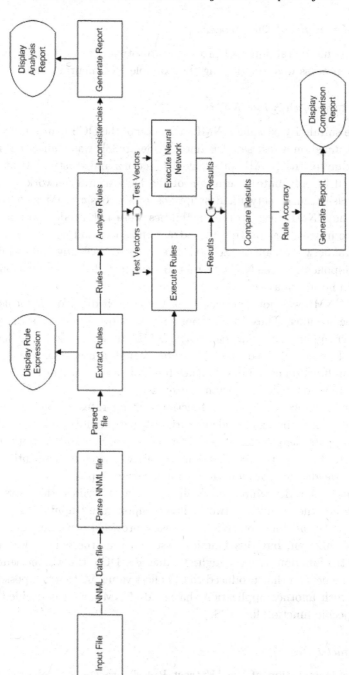

Fig. 5.   General neural network rule extraction tool process.

3.3. *An Example of the Process*

To test the rule extraction tool process a prototype was developed. Each step of the process was tested using the example DCS neural network.

3.3.1. *Translating DCS into NNML*

To test the suitability of using NNML, a version of the DCS neural network that was trained on a test set (Iris data [Fisher 1936]) was translated into NNML. Many of the NNML pre-defined functions and existing structure tags allowed for adequate description of the DCS neural network. These included basic neural network descriptions and a basic SOM structure. Some of the NNML description capabilities were not needed, primarily because the interest was on the rule extraction aspect and not the history of the neural network training. Sections like the pre-processing and training process components to the NNML structure were included for compatibility but contain no information.

While NNML was not a perfect fit for the DCS, it did provide for user-defined declarations. These declarations must follow a certain structure giving other applications that can read NNML the ability to parse these functions. Two user-defined functions were created: a user-defined neuron combination function called dcs_distance function, and a user-defined post-processing function called dcs_winner_takes_all.

The dcs_distance function is a function that operates on each neuron. It takes as input the independent variables (sepal_width, sepal_length, petal_width, petal_length) and the weights of the DCS neural network. For this project, the dcs_distance function will allow adequate description of the neuron weights and provide the tool a necessary input.

Similarly, the dcs_winner_takes_all function describes the possible outputs from the neural network based upon the inputs into the dcs_winner_takes_all function. NNML already provided a winner_takes_all pre-defined function, but this function assumed that the winner was the input into the function with the highest value. For DCS, the winner neuron would be the neuron that produced the smallest value. So to avoid possible confusion with another application that reads NNML, it was decided to create a specific function for DCS.

3.3.2. *Extract Rules*

For the implementation of the "Extract Rules" process, two algorithms, written in Matlab, were employed. The original algorithm (Fig. 1) that generates human readable rules [Darrah 2004] and the second algorithm (Fig. 3) that generates deterministic rules were implemented and tested. This second algorithm was also used in a two-step process to help refine the rules generated by the original algorithm as described in a previous section.

The new deterministic rule extraction algorithm was first tested on a general two-dimensional example that could be hand checked for accuracy. The following points were used as the centroid (weight) set, $C = \{(2, 3), (5, 7), (9, 5), (7, 1), (12, 3), (11, 7), (14, 5)\}$. The connection matrix was determined by hand. A DCS neural network was set up with these parameters. The deterministic algorithm was used to develop the rules that were implemented. Test data was used to generate output from both the rules and the DCS. The output from the deterministic rule set had 100% agreement with the DCS output on 50,000 randomly generated test points.

After testing the algorithm on the general example given above, the DCS neural network was trained on the Iris benchmarking data set. The Iris data set has four independent variables and is widely used to test different algorithms. The set is interesting because it is not linearly separable.

After training the DCS on the Iris data, rules were extracted by applying the algorithm to the weights and connection matrix. At this time various disagreements were found with the rule output and the DCS output. Through investigation, it was found that since DCS is a topology-preserving structure, certain connections are purposely severed during adaptation if subsets of the input space are disjoint. This caused a slight discrepancy with the way DCS operates and the way the rules were developed assuming full connection based on geometric positioning of the weights. This discrepancy was overcome by using the original connection matrix generated by the DCS (not the geometric connection matrix created by a Delaunay triangulation) and breaking the output into four different cases: (1) weight has no neighbor, (2) weight has one neighbor, (3) weight has two neighbors, and (4) weight has three or more neighbors. In each case, the rules must be generated in a slightly different way for those regions in order to reach complete agreement with the DCS output. After this modification to the rule extraction code, the rule output and the DCS output had 100% agreement on the entire Iris original data set and on 50,000 randomly generated points.

The last test of the rule extraction algorithm was on the IFCS flight data. The flight data used was obtained from an F-15 Flight Simulator developed at West Virginia University for the IFCS. The data set used has independent variables: mach, altitude, alpha, and beta, and dependent variables cza, czdc, czds. The DCS was trained and the rules were extracted by applying the algorithm to the weights and connection matrix. The rule output and the DCS output had 100% agreement on the entire original data set and on 50,000 randomly generated points.

3.3.3. *Analyze Rules*

This process block "Analyze Rules" represents the need to implement an automatic formal analysis of the extracted rules as a representative of the neural network. Several alternatives for this process have been identified and are discussed in this section. In addition to the determination of test vectors previously indicated as output, it is envisioned that this process will generate a report that contains other information including whether mathematically expressed assertions representing requirements are true, false, or not provable.

After the rules are generated, they can be used to help produce test vectors for each region of the input domain and possible system output. The rules define the regions of the input space that correspond to a specific output function. Once the input domain is partitioned into regions by the rules, test data for each of these regions and on or near the boundaries between regions can be generated. This will ensure that test sets have complete coverage of the input domain and can be used to determine if there will be smooth transitions between neighboring regions.

An effort to identify commercial-off-the-shelf analysis tools needed for additional analysis of the rules led in several different directions. It was determined that current model analysis tools such as SCRTool or SPIN are not able to analyze models that are not fundamentally based on discrete state machines. Constance Heitmeyer, the head of the Software Engineering Section of the Naval Research Laboratory's Center for High Assurance Computer Systems that develops SCRTool, affirmed this and stated that a theorem prover would be a more likely candidate for such an analysis stage [Heitmeyer 2004].

Research into the current state-of-the-art of theorem provers showed promise that a semi-automated approach to analyzing the rules could be provided. One such theorem prover, Salsa, also developed by the Naval Research Laboratory, will likely become part of the SCRTool suite [Bharadwaj 2000]. The rules extracted from the neural network can be used to build a model of the system for input to the Salsa theorem prover.

Violations can be stated as invariants to be checked against the system model. Unlike model checkers, Salsa can handle infinite state spaces through use of induction and symbolic encoding of expressions involving linear constraints.

Theorem provers are an appropriate long-term solution, however, other solutions, such as geometric analysis, may be more immediately effective. Since the DCS deterministic rules give a complete geometric model of the trained neural network, assertions can be formulated geometrically as a region or hyperplane of the input space. These assertions could be stated in the form of "when this condition is met by the inputs, the outputs must not be in this region". The assertions can then be checked geometrically to verify if they will intersect with other regions of the input space to determine if violations can occur.

Geometric analysis of the neural network rules can also be done without the guidance of explicit requirements. The antecedents of the rules geometrically describe the partition of the input domain. These regions can be inspected for size and percent coverage of the domain. This geometric approach can take advantage of the vast amount of previous research in computational geometry and computer graphics algorithms and hardware.

4. Verification and Validation Examples

The following scenarios show how extracted rules can help verify and validate a neural network. The neural network in the scenarios is a Matlab implementation of the DCS neural network that was has put together for testing purposes. This neural network has the same characteristics as the DCS that was implemented in the IFCS mentioned earlier.

Scenario 1: Human Understandable Rules Led to Identification of Coding Error.

The original rule extraction algorithm, which generated human understandable rules, is based on how an input stimulus is matched to a centroid of the DCS or its best matching unit (BMU). The human understandable rules support verification inspection methods. Each input stimulus results in the selection of a BMU internal to the DCS network. The BMU is considered the centroid of a cell and each input that related to that BMU is considered to be a member of that cell. The human understandable rules were generated to describe each cell. The minimum and maximums of each input variable related to a specific BMU are used in the rule antecedents. The minimum and maximum of each output variable associated with this cell make up the rule consequent. Any BMU that did not match input stimulus did not generate a rule. (See rule algorithm.)

When the human understandable rule algorithm was applied to a DCS network that had been trained on the Iris data, a discrepancy was noted between the number of rules generated and the number of nodes within the DCS network. There were fewer rules than nodes. This implied that for the set of input data used to train the neural network, a node was established that never matched any of the other data, and thus these BMUs did not have corresponding rules. This led to investigation of the existence of these nodes by walking through the source code and looking for problems. Debugging and execution traces pointed to a problem in some of the DCS code that had been optimized to run within a Matlab environment. The original IFCS DCS code was developed within the C programming language. For optimization purposes, when the code was moved into a Matlab script for experimentation, all usage of "for" loops were removed and replaced with vectorized math. One of the lines of code used for the optimization dealt with the identification of BMUs, and incorrectly referenced the BMU variable.

Instead of only looking across the existing set of nodes within the DCS network, it made use of the DCS maximum allowed number of nodes. In effect, when looking for the BMU, the DCS was allowed to consider nodes which had not yet been assigned, and by default were at zero value and can be thought of as existing at the origin. At times, these nodes were actually better at matching the input than any one of the existing nodes, and DCS manipulated these non-assigned nodes when it should not have. The result was that nodes that had not been assigned learned and adapted. They showed up as having non-zero values and appeared to be nodes upon visual inspection of the DCS structure, but did not actually exist. DCS was losing some potential learning within these nodes. The rules ignored these nodes since they were not able to ever become BMUs that led to the discrepancy. The line of code was modified to ignore non-assigned nodes, and then DCS nodes correctly matched up with the human understandable rules.

Scenario 2: Machine Understandable Rules Led to Identification of Two Coding Errors.

The deterministic rule extraction process is designed to have 100% agreement with the performance of the DCS network. However, testing of some of the first sets of deterministic rules showed that there was a large disagreement between the rules and DCS.

The rules were re-structured so that the antecedents were broken into a rule pertaining to each BMU, and then under the BMU rules, each neighboring SEC rule was present. This allowed comparison to see if the errors between the rules and DCS were based upon BMU selection, SEC selection, or within the consequent. By comparing the BMU output from DCS with the specific BMU rule corresponding to the input, it was discovered that the BMU selection was consistent between the rules and the DCS. But the selection of the SEC was not matching between the two. Further investigation required analysis of the DCS recall function.

In the DCS recall function, two errors within the same line of code were discovered. One was related to substitution of the "max" for "min" commands within DCS. For the recall function to perform properly, the smallest distance is always used to identify the closest node to a stimulus. This is true also when selecting the second closest node from among a BMU's neighbors. But the code showed that the "max" function was being used in place of the "min" function. This would subsequently show up within the DCS recall function as the DCS always selected the node furthest away from the stimulus.

Figure 5 shows how the DCS, trained on the example data, assigned a BMU to random data. The different colors correspond to the different centroids of a Voronoi region that could be chosen as a BMU. If a data point was assigned centroid one as its BMU then the data point would be plotted with the color of centroid one. Figure 6 below shows that with the code errors the boundaries were not precise and there was incorrect overlap of regions. The DCS neural network was assigning the incorrect BMU to the data points.

Further, this same line of code contained an incorrect reference to the strengths of the neighborhood for the BMU rather than the distances of the neighbors from the stimulus. This mistake is quite a large mistake, but due to the nature of the small DCS networks, and the small values on which the network was learning, the mistake was masked much of the time. Normal testing of the DCS showed that it could achieve accuracies above 90%, even with this error present. The robustness of the DCS network made discovery of this error difficult.

After the coding errors were corrected, random data was again plotted. Figure 7 shows the data is now partitioned into distinct regions and these

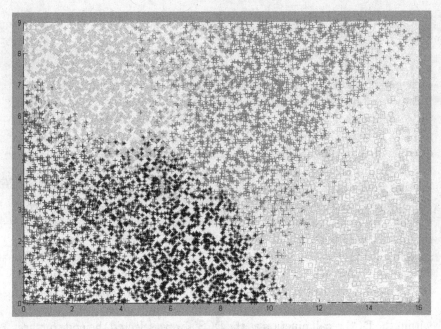

Fig. 6.   Sample data plotted with code error.

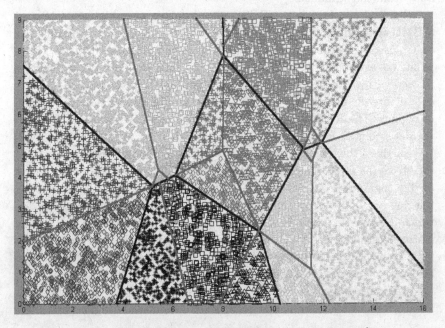

Fig. 7.   Sample data plotted after code errors were corrected.

regions completely match the boundaries in the rule antecedents as pictured in Fig. 4.

The line of code was changed to consider the distances rather than the connection strengths and to choose the min instead of max. The rules and DCS were compared again. This gave the expected results of 100% agreement. The deterministic rules were deemed a success because they had allowed for the discovery of two coding errors, which were not readily apparent during normal testing.

5. Potential Applications

5.1. *Certification of Neural Networks*

Techniques applied towards the certification of fixed neural networks may not provide enough confidence to allow their use in a high-assurance system. One way to remove that concern is the complete replacement of the neural network with the deterministic rules. The rules become an expert system. In the case of the DCS neural network, the implemented set of rules describing the behavior of the neural network system matched with 100% agreement. Verification and Validation of expert systems is a mature field, and expert systems have been used in high-assurance systems before including the NASA Deep Space 1 exploration mission as an AI planner. The rule extraction provides for the translation of a relatively new field, neural networks, into a relatively experienced field, expert systems.

5.2. *Health and Status Monitoring of the Neural Network*

Neural network rules can assist in the creation of safety monitors to judge the correctness of a neural network system.

A safety monitor is a device (software or hardware) that exists external to the neural network within a safety- or mission-critical system. Its role is to judge the neural network for stability, convergence, correctness, accuracy, or any number of other factors to determine in real-time if the system is operating within acceptable boundaries. For the IFCS systems under development by NASA DFRC, safety monitors have been developed to observe both the fixed and the dynamic neural networks.

In both fixed and dynamic neural network cases for the IFCS program, the design of a safety monitor has been a somewhat complex and difficult process. For example, the safety monitor used for the pre-trained neural network within the first generation of the IFCS program was a set of error-code like correction equations that were not easily understood or explainable. The purpose use of these equations was to determine if the output from the pre-trained neural network was within some allowed percentage variation. Neural network rules, extracted from the fixed, pre-trained neural network would have served the same purpose, been far easier to calculate, and would have had a direct relation to the underlying knowledge embedded in the neural network.

The same could be said for a dynamic neural network. Rule extraction conducted in real-time would provide a way to compare against a set of high-level requirements to determine if the neural network operates correctly.

Consider, for example, the use of extracted human understandable rules. These rules contain expected input-output ranges. Much of the IFCS software operates with acceptable range limits, or robustness measures, which describe system safety limits for input or output values. Human understandable rules can be used to determine if the network will operate within these limits. Ranges that exceed these limits can be caught by a simple comparison against the safety requirements and the neural network system can be flagged as incorrect.

Another approach would be to use the deterministic rules that provide the same output as the neural network. For the research conducted within NNRules, the only information needed to construct these deterministic rules for the DCS are the weights and the connection matrix. This information is easily communicated to an external system from the neural network where the rules can be extracted, and then used to judge the correctness of the output from the DCS. The rules then become an external representation of the DCS, independent of the DCS source code and the specific instance of DCS in memory, and able to replicate the way the DCS would behave. Careful design of the rules-safety-monitor approach provides a way to judge if the data output from the DCS was corrupted by the transmission medium, and a way to judge if the internal knowledge of the DCS violates a safety requirement.

5.3. *Extracted Rules as Basis for Expert Systems*

Neural networks can be useful for fault diagnosis. Fault diagnosis using neural networks has the same structure as model-based methods: a set of signals, carrying fault information is fed into the neural network which outputs a vector fault signal indicating normal and faulty system operation

[Pouliezos 2002]. Neural network diagnosis is aimed at overcoming the shortcomings of process model-based techniques that require mathematical models that represent the real process satisfactorily. These neural network fault diagnosis techniques can be seen as a Pattern Recognition problem involving three stages: measurement, feature discovery, and classification.

Although neural networks are powerful data mining tools for analyzing and finding patterns in data, they represent a class of systems that do not fit into the traditional paradigms of software development and certification. Instead of being programmed, a learning algorithm "teaches" a neural network using a representative set of data. Often, because of the non-deterministic result of the "learning", the neural network is considered a black box and its response may not be predictable. This concern limits the use of neural networks in high assurance systems, such as safety- or mission-critical software systems.

Rule extraction enables the knowledge gained by the neural network during training to be expressed as a deterministic rule set. Neural network rule extraction techniques translate the decision process of a trained neural network into an equivalent decision process represented as a set of rules. The rules can be used to build an equivalent expert system that would have predictable behavior.

The following steps can be used to create a fault monitor for fault diagnosis in a safety-critical system in which a neural network could not be included because of restrictions.

1. Perform data mining using the neural network to obtain failure mode patterns or failure type classifications.
2. Extract rules from neural network to build knowledge base of potential failures.
3. Create prognostic algorithms using the results of the rule extraction.
4. Implement prognostic algorithms as an expert system.

6. Conclusion

The NASA, Department of Defense (DOD), and FAA are currently researching the potential of neural networks and how they can enhance current technology. Given the adaptive and non-deterministic characteristics of neural networks, it has been difficult to provide trust in these systems. These agencies have spent years researching the functionality of neural networks and have used "brute force" testing to determine if these intelligent systems are operating as expected. As neural networks are becoming more common in critical systems, there is a growing need to verify and validate them beyond "brute force" testing. The technology developed under this research enables, through the use of rule extraction, the use of additional verification and validation methods that contribute toward achieving the levels of trust needed to deploy neural net based systems. Extracting the rules and/or knowledge of the neural network may help provide that necessary level of trust.

NASA's interest in neural networks stems from their plans for deep space missions of the future (Mars and beyond). In order for these missions to be successful, the deployed spacecraft must be able to adapt to unknown or unforeseeable situations. The learning power of neural networks provides that adaptability.

The DOD, specifically the Navy, has been researching how to incorporate neural networks into surface ship and submarine operations. In 1997, the Navy ran successful tests on a prototype "Smart Ship" (USS Yorktown) that was designed to use various monitoring systems to assist in ship operations.[4] The technology introduced on the Smart Ship was designed to minimize the size of the crew and the workload on the remaining crew while still maintaining the high level of readiness. As an enhancement to the Smart Ship technology, the Navy is turning towards neural networks.

The FAA's interest in neural networks is actually a joint effort with NASA Ames Research Center. They are researching how neural networks can be used in flight control systems. With the successful tests of neural networks in critical systems on the Dryden F-15 test plane, the FAA is interested in seeing how this technology could be used in commercial carriers. With the recent tragedy of 9/11, neural network technology has the potential of providing safer flight with intelligent planes. The neural networks learn what the flight envelopes are and will be trained not to fly outside of those envelopes. The research in this area is far from hitting the commercial carriers, but much progress has been made.

Although these research efforts may not be implemented for many years, the question of verification and validation of neural networks must be answered now. Extracting the rules and knowledge of a neural network is one of many steps that can be used to verify and validate the system, as well as provide a level of trust in the system.

---

[4]http://www.fcw.com/fcw/articles/1997/FCW_100697_1131.asp

Acknowledgements

This research was sponsored by NASA Ames Research Center, under Research Grant NAG5-12069.

References

1. Andrews, R., Diederich, J., Tickle, A.B.: A survey and critique of techniques for extracting rules from trained artificial neural networks, *Knowledge Based Systems* 8, 1995a, 373–389.
2. Andrews, R., Geva, S.: RULEX & CEBP networks as the basis for a rule refinement system. In Hybrid Problems, Hybrid Solutions, ed. John Hallam. IOS Press, 1995b.
3. Andrews, R., Geva, S.: Rule extraction from local cluster neural nets, *Neurocomputing* 47, 2002, 1–20.
4. Bharadwaj, Ramesh, Steve Sims: Salsa: Combining constraint solvers with BDDs for automatic invariant checking. In *Proc. Tools and Algorithms for the Construction and Analysis of Systems* (TACAS 2000). Lecture Notes in Computer Science, Springer, 2000.
5. Bruske, J., Gerald, S.: Dynamic cell structures. In *Proceedings of Neural Information Processing Systems* (NIPS), 1994, 497–504.
6. Carpenter, A., Tan, A.H.: Rule extraction: from neural architecture to symbolic representation, *Connection Science* 7(1), 1995, 3–27.
7. Craven, M.W., Shavlik, J.W.: Using sampling and queries to extract rules from trained neural networks' Machine Learning, *Proceedings of the Eleventh International Conference,* San Francisco CA, 1994.
8. Darrah, Marjorie and Brian Taylor: Rule extraction from dynamic cell structure neural network used in a safety critical application. In *Proceeding of Florida Artificial Intelligence Research Society Conference,* Miami FL, May 2004, 17–19.
9. Fisher, A.: *Annals of Eugenics* 7, 1936, 179–188.
10. Filer, R., Sethi, I., Austin, J.: A comparison between two rule extraction methods for continuous input data. In *Proc. NIPS'97 Rule Extraction from Trained Artificial Neural Networks Workshop,* Queensland Univ. Technology, 1996.
11. Fritzke, B.: Growing cell-structures — A self-organizing network for unsupervised and supervised learning, *Neural Networks,* 7(9), 1994, 1441–1460.
12. Fu, L.M.: Rule generation from neural networks, *IEEE Transactions on Systems, Man, and Cybernetics,* 28(8), 1994, 1114–1124.
13. Giles C.L., Omlin, C.W.: Extraction, insertion, and refinement of symbolic rules in dynamically driven recurrent networks, *Connection Scientist,* 5, 1993, 307–328.
14. Goddard N., Hucka, M., Howell, F., Cornelis, H., Skankar, K., Beeman, D.: Towards NeuroML: Model description methods for collaborative modeling in neuroscience, *Phil. Trans. Royal Society,* Series B, 2001, 356–1412, 1209–1228.

15. Heitmeyer, C.: Personal communication with Rhett Livingston, 2004.
16. Johansson, Ulf and Rikard Konig: The truth is in there — Rule extraction for opaque models using genetic programming. In *Proceeding of Florida Artificial Intelligence Research Society Conference*, Miami FL, May 2004, 17–19.
17. Krishnan, R.: A systematic method for decompositional rule extraction from neural networks. In *Proc. NIPS'97 Rule Extraction from Trained Artificial Neural Networks Workshop*, Queensland Univ. Technology, 1996.
18. Martinetz, T.M.: Competitive Hebbian learning rule forms perfectly topology preserving maps. In *Proceedings of International Conference on Artificial Neural Networks* (ICANN), 1993, 427–434. Amsterdam: Springer.
19. McMillan, C., Mozer, M.C., Smolensky, P.: The connectionist scientist game: Rule extraction and refinement in a neural network, *Proc. of the Thirteenth Annual Conference of the Cognitive Science Society*, 1991.
20. Optiz, D.W., Shavlik, J.W.: Dynamically adding symbolically meaningful nodes to knowledge-based neural networks, *Knowledge-Based Systems*, 8, 1995, 301–311.
21. Robtsov, D.V., Sergey, V.B.: Neural Network Markup Language, NNML Home Page [online]. [cited June 15, 2004]. Available on the World Wide Web: ⟨http://www.nnml.alt.ru/index.shtml⟩, 2004.
22. Saito, K., Nakano, R.: Law discovery using neural networks. In *Proc. NIPS'96 Rule Extraction from Trained Artificial Neural Networks Workshop*, Queensland Univ. Technol., 1996.
23. Schellhammer, I., Diederich, J., Towsey, M., Brugman, C.: Knowledge Extraction and Recurrent Neural Networks: An Analysis of an Elman Network Trained on a Natural Language Learning Task. QUT-NRC Tech. Rep. 97-IS1, 1997.
24. Sestito, S., Dillon, T.: The Use of Sub-symbolic Methods for the Automation of Knowledge Acquisition for expert Systems. In *Proc. 11th International Conference on Expert Systems and their Applications* (AVIGNON'91), 1991, Avignon, France.
25. Sestito, S., Dillon, T.: Automated knowledge acquisition of rules with continuously valued attributes. In *Proc. 12th International Conference on Expert Systems and their Applications* (AVIGNON'92), 1992, Avignon, France.
26. Sestito, S., Dillon, T.: *Automated Knowledge Acquisition*, Prentice Hall: Australia, 1994.
27. Setiono, R.: Extracting rules from neural networks by pruning and hidden unit splitting, *Neural Computing*, 9, 1997, 205–225.
28. Tickle, A.B., Orlowski, M., Diederich, J.: DEDEC: Decision Detection by Rule Extraction from Neural Networks. *QUT NRC*, 1994.
29. Tan A.W.: Cascade ARTMAP: Integrating Neural Computation and Symbolic Knowledge Processing. In *IEEE Trans. Neural Networks*, 8, 1997, 23–250.
30. Tickle, A.B., Andrews, R., Golea, M., Diederich, J.: The truth is in there: Directions and challenges in extracting rules from trained artificial neural networks, 1998.

31. Towell, G.G., Jude, W.S.: Extracting refined rules from knowledge-based neural networks. *Machine Learning*, 13(1), 1993a, 71–101.

32. Towell, G., Shavlik, J.: The extraction of refined rules from knowledge-based neural networks. *Machine Learning*, 13(1), 1993b, 71–101.

33. Thrun, S.: Extracting rules from artificial neural networks with distributed representations. In *Advances in Neural Information Processing Systems (NIPS) 7*, eds. Tesauro, G., Touretzky, D., Leen, T. Cambridge, MA: MIT Press, 1995.

34. Viktor, H.L., Engelbrecht, A.P., Cloete, I.: Reduction of Symbolic Rules for Artificial Neural Networks Using Sensitivity Analysis. *IEEE Conference on Neural Networks*, Perth, Western Australia. November, 1995.

Appendix A

Rule extraction algorithm	Rule translucency	NN portability	Rule format
DCSRules [Darrah 2004]	Decompositional	Applies only to DCS networks, may be applicable to SOMs	Propositional if...then...else
KT Rule Extraction [Fu 1994}	Decompositional	Neural Networks utilizing standard back-propagation as the training regime.	*if* $0 \leq output \leq threshold\_1 \Rightarrow no$; *if* $threshold\_2 \leq output \leq 1 \Rightarrow yes$; $threshold\_1 < threshold\_2$
Subset Algorithm [Towell 1993a, 1993b]	Decompositional	Described as being "general purpose"? May only be for NNs with neurons that are binary output	Propositional if...then...else
Validity Interval Analysis [Thrun 1994]	Pedagogical	Maybe only feedforward NNs	Propositional if...then...else
Rule-Extraction-As-Learning [Craven 1994]	Pedagogical	Seems to work outside of the specified NN by learning the training set according to the output of the NN	Propositional if...then...else and M-of-N

(Continued)

Rule extraction algorithm	Rule translucency	NN portability	Rule format
DEDEC [Tickle 1994]	Eclectic/ Pedagogical	The only constraint on the ANN is that it be a multilayer, feedforward ANN trained by the "back-propagation" method.	Symbolic rules akin to: if *variable* in {...} then ...
KBANN/M-of-N [Towell 1993a, 1993b]	Decompositional	Specific to the KBANN NN architecture	(modified) Boolean [If (M of the following N antecedents are true) then ....]
BRAINNE [Sestito 1991, 1992, 1994]	Pedagogical	Designed to extract rules from a NN trained using the back-propagation technique.	Propositional if ... then ... else
Fuzzy ARTMAP [Carpenter 1995]	Decompositional	Algorithm only works on Fuzzy ARTMAPs	Fuzzy rules (no examples)
Sensitivity Analysis Algorithm [Viktor 1995]	Decompositional	Unknown	Unknown
COMBO Algorithm [Krishnan 1996]	Decompositional	Generally applicable to feedforward networks with Boolean inputs	Propositional with Boolean input IF $\Sigma w_{pm}+$ bias on neuron $>$ threshold on neuron THEN concept corresponding to the neuron is true

(Continued)

Rule extraction algorithm	Rule translucency	NN portability	Rule format
RULEX [Andrews 1995a, 1995b, 2002]	Decompositional	The constrained error back-propagation (CEBP) multilayer perceptron (similar to radial basis functions)	Propositional if...then...else
RF5 [Saito 1996]	Decompositional	Specific NN architecture where the ANN uses "product units" in the hidden layer (power units)	Scientific Laws $y_t = c_0 + \sum_{i=1}^{h} c_i x_{t1}^{w_{i1}} \cdots x_{tn}^{w_{in}}$
Knowledge Extraction and Recurrent Neural Networks [Schellhammer 1997]	Decompositional	Generally applicable to recurrent neural networks with Boolean input	Deterministic Finite-State Automaton (DFA)/State Transition rules
TopGen [Opitz 1995]	Decompositional	Specialized neural network architecture, specialized training algorithm	Propositional/ M-of-N rules
RX [Setiono 1997]	Decompositional	Generally applicable to supervised feedforward networks constrained to a single hidden layer	Propositional
Interval Analysis Algorithm [Filer 1996]	Pedagogical	Multilayer perceptrons	Propositional with continuous or discrete input IF $\forall\, 1 \leq i \leq n$: $x_i \in [a_i^{min}, a_i^{max}]$ THEN concept represented by the unit is true

(*Continued*)

Rule extraction algorithm	Rule translucency	NN portability	Rule format
Extraction of Finite-State Automata from Recurrent Network [Giles 1993]	Decompositional	Portable (may only be usable with recurrent neural networks)	DFA/State transition rules
TREPAN [Craven 1994]	Pedagogical	Suitable for any trained/learned model	Decision Tree with M-of-N split tests at the nodes
Cascade ARTMAP [Tan 1997]	Decompositional	Relies on specific architecture, cascade ARTMAP	Propositional with discrete input
RuleNet & The Connectionist Scientist Game [McMillian 1991]	Decompositional	Uses a specific type of neural network architecture: RuleNet	Propositional if...then...else (the authors use the term "condition-action" rules)

Chapter 6

REQUIREMENTS ENGINEERING VIA LYAPUNOV ANALYSIS FOR ADAPTIVE FLIGHT CONTROL SYSTEMS

GIAMPIERO CAMPA

Mathworks, 400 Continental Blvd, Suite 600,
El Segundo, CA, 90245, USA
Giampiero.Campa@mathworks.com

MARCO MAMMARELLA

Advanced Space System and Technologies Business Unit,
GMV Aerospace and Defence S.A.
Isaac Newton, 11 P.T.M. Tres Cantos, 28760, Madrid
marco__mm@hotmail.com

MARIO L. FRAVOLINI

Department of Electronics and Information Engineering,
University of Perugia, 06100 Perugia, Italy.
fravolin@diei.unipg.it

BOJAN CUKIC

Lane Department of Computer Science and Electrical Engineering,
West Virginia University, Morgantown, WV 26506/6106
Bojan.Cukic@mail.wvu.edu

Adaptation creates new challenges for software and system requirements engineers. If the system is expected to change its operation, how do we express and ensure the bounds of "desirable" changes, i.e., those that by some measure improve rather than deteriorate the performance? This paper introduces a technique for the analysis of boundedness problem for a general class of dynamical systems subjected to direct adaptive control using Lyapunov analysis, within an "adaptive augmentation" setting. An accurate formulation of the bounding set for the overall closed loop system is derived, under the assumption that the error driving the adaptive element is chosen as a static function (i.e., a matrix gain) of all the measurable outputs of the system. Based on this result, we present a simulation example relative to an adaptive flight control system designed for F/A-18 aircraft. Further, we discuss the impact that typical design parameters have on the shape and dimensions of the bounding set.

1. Introduction

Adaptive control systems could allow future aircraft and spacecraft to cope with different unanticipated component failures, while at the same time exhibiting improved performance and autonomy under uncertain environmental conditions. However, a fundamental concern with adaptive systems is that their evolution may lead to undesirable or unsafe levels of performance, and ultimately, to unacceptable behavior. This is especially a problem with safety critical applications, where in fact the control system must incorporate a rigorous validation and verification process to comply with certification standards. In particular, the very first phase of the validation process, the requirements validation, represents novel challenges. This happens because, due to its intrinsic nature, an adaptive system can alter its behavior in response to changing environmental conditions, and it is therefore challenging to place sensible requirements if the system can substantially evolve itself in the field, following the deployment.

Theoretically, the problem of formulating verifiable requirements for adaptive systems could be addressed using results from "boundedness" proofs based on Lyapunov analysis. These proofs typically show that the evolution of the adaptive system is Uniformly Ultimately Bounded (UUB) into a certain compact set (also known as the UUB set). Calculating the UUB set could give the system designer the capability to immediately check whether such bounding set intersects regions of the state space where the behavior is not guaranteed to be safe, with implications to software and system verification and validation goals. In practice however, the evaluation of the UUB set for an adaptively controlled system is a rather long and cumbersome process. For this reason, to the best of our knowledge, it has never been carried out in the literature at the level of completeness which would satisfy thorough validation of adaptation requirements.

Furthermore, due to the necessity of reducing length and complexity, a great number of limiting assumptions, simplifications and overestimations are typically made within the proof, resulting in two main shortcomings. First, the resulting estimation of the bounding set is overly conservative and may tend to infinity,[1] therefore defeating the validation purpose. Second, there is a considerable lack of generality due to the simplifying assumptions that are made on the structure of the system.[8] In other words, applications for which such proofs are derived are significantly simpler than those that commonly appear in practice. Recent advancements[17,20] remove some limitations by adopting a more general "adaptive augmentation" viewpoint, in which an adaptive element is added to a (possibly pre-existing) dynamical system to partially cancel the unwanted effect of an uncertainty term, which may be due to perturbations, failures and/or poorly modeled dynamics. Nevertheless, the other shortcomings remain.

In this chapter, the boundedness problem will be considered in the general framework of static output feedback adaptive augmentation for a Multiple Input Multiple Output (MIMO) plant. Only a limited number of assumptions will be made on the system to be augmented, therefore leading to a general treatment encompassing a wide variety of adaptive control systems, which will in turn yield more general and less conservative expressions for the system's bounding sets. On the basis of the obtained expressions, the conditions under which boundedness can be guaranteed will be analyzed in detail, and the interactions between the design parameters and the shape and dimensions on the bounding sets will be discussed.

The chapter is organized as follows. The framework for the adaptive augmentation problem is overviewed in the Section 2. Section 3 describes the Lyapunov analysis, followed by the characterization of the bounding set. Section 4 is devoted to an illustrating example in which the bounding set is calculated for the adaptively controlled linear dynamics of the F/A-18 aircraft.

2. The Framework for Adaptive Augmentation

2.1. *The Plant, the Closed Loop System and the Error Dynamics*

A nonlinear, uncertain, time-invariant, continuous time, MIMO dynamical system to be controlled (also referred to, in control engineering parlance, as "the plant") can be described by the following equations:

$$\dot{x}(t) = Ax(t) + Bu(t) + \Delta_x(x(t), u(t))$$
$$y(t) = Cx(t) + Du(t) + \Delta_y(x(t), u(t)) \tag{1}$$

where the matrices $A \in \mathbb{R}^{n \times n}$, $B \in \mathbb{R}^{n \times m}$, $C \in \mathbb{R}^{v \times m}$, $D \in \mathbb{R}^{n \times m}$ are known and constant, and the "uncertainty" terms $\Delta_x(x(t), u(t)) \in \mathbb{R}^{n \times 1}$ and $\Delta_y(x(t), u(t)) \in \mathbb{R}^{v \times 1}$ are continuous globally and uniformly bounded *unknown* functions. Equations (1) can accurately model a wide variety of physical systems and processes whose behavior is governed by differential equations.[8] Examples include mechanical, electrical, electronic, thermal, fluid-hydraulic,[9] as well as chemical,[10] biological,[17] ecological,[18] economical[11] and quantum-mechanical[12] systems. The vector of *output* variables $y(t)$ contains the available measurements collected from the system at time t, while the vector of *input* variables $u(t)$ contains the commands imposed by the environment (which can include an external device or a user), on the system. The vector of *state* variables $x(t)$ contains the variables describing the internal state of the system, which are typically not readily measurable. The two uncertainty terms include the unknown terms that cannot be captured by a simple linear description of the system.

As mentioned earlier, additional assumptions are typically made on the system described by Eq. (1) with various degrees of generality losses. Earlier research relied on $y(t)$ and $u(t)$ to be scalars (SISO assumption) or on the system to be feedback linearizable, while more recent research relied on C being the identity matrix (state feedback assumption), on $C = B^T P$, (SPR assumption) or on Δ_x belonging to the space generated by the columns of B. Also, in the majority of cases, both D and Δ_y are assumed to be zero. In this chapter, all the above assumptions, except the boundedness of both Δ_x and Δ_y and the observability of the system of Eq. (1), will be relaxed.

2.2. *The Linear Controller, the Closed Loop System and the Error Dynamics*

The plant is typically connected to a "controller", which is typically a microprocessor-based system designed so that the *closed loop* system — that is the overall system containing both the physical system Eq. (1) and the controller — is stable and has an output $y(t)$ that tracks a certain desired reference signal $r(t)$. Graphically, the closed loop system is as shown in Fig. 1, where the underscore symbol indicates a subscript notation, and where $u_{ad}(t)$ is the contribution specified by a (yet to be described) adaptive element.

The derivation of the dynamics of the closed loop system is achieved following standard rules and definitions. By appropriately defining the matrices A_z, C_z, B_u, B_y, and D_y, as a function of the previously defined

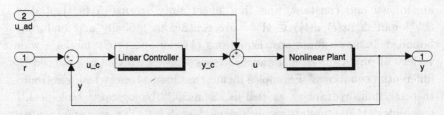

Fig. 1.   Closed loop system.

matrices from Eq. (1) the "error dynamics" can be written as following:

$$\dot{z} = A_z z + B_u u_{ad} + B_y \Delta_y + B_x \Delta_x$$
$$e_y = C_z z + D_u u_{ad} + D_y \Delta_y \tag{2}$$

The above dynamical system represents the dynamics of the error (that is, the difference) between a "reference" system — which models the desired behavior — and the actual system. Therefore, keeping the actual system behavior as similar as possible to the desired behavior is equivalent to keeping the variable z in Eq. (2) as close as possible to zero. It is important to notice that the error dynamics in Eq. (2) is asymptotically stable, in the sense that $z(t)$ and $e_y(t)$ asymptotically approach zero whenever the forcing signal $B_u u_{ad} + B_x \Delta_x(x, u) + B_y \Delta_y(x, u)$ is zero. Therefore, the goal of the adaptive system is to learn and compensate for the uncertainties, so that the forcing signal is kept as small as possible.

2.3. *The Uncertainty*

The uncertainties can be divided into the sum of two contributions, specifically the *matched* and *unmatched* uncertainty vectors.[21] The matched uncertainty vector Δ can be interpreted as a disturbance having the same input matrix as the input command. If this uncertainty was known, it could be cancelled by an opposite command signal. The remaining parts of the total uncertainty are the unmatched vectors Δ_s, and Δ_o, where the subscripts "s" and "o" indicate the state and the output equation respectively. Rewriting Eq. (2), in terms of the matched and unmatched uncertainties yields:

$$\dot{z} = A_z z + B_u(\Delta + u_{nn}) + \Delta_s$$
$$e_y = C_z z + D_u(\Delta + u_{nn}) + \Delta_o \tag{3}$$

Using standard arguments from finite-difference theory, assuming the observability of the system in Eq. (2), and appropriately defining the vector η, which contains all the known information on the overall system, leads to:

$$\Delta = \Delta'(\eta) + \varepsilon_1(\eta) \tag{4}$$

In words, Eq. (4) states that the unknown Δ can be reconstructed with an arbitrary small error $\varepsilon_1(\eta)$ from a sufficient number of input and output samples by selecting a sufficiently small sampling time. The nonlinear function $\Delta\prime(\eta)$ can then be approximated (or, in other words, learned) by the adaptive element in the control loop.

2.4. *The Adaptive Element*

Due to their simplicity and their approximation capabilities, neural networks such as Radial Basis Function (RBF) networks,[3] Single Hidden Layer sigmoidal networks,[4] and a fully-adaptive growing RBF (GRBF) networks,[5] are frequently used as adaptive elements in a great number of adaptive control applications. The typical adaptation algorithm for an RBF network used for control purposes can be expressed as following:

$$\dot{\hat{W}}(t) = Proj(\hat{W}(t), L\Phi(\eta(t))e_{nn}^T(t) - \gamma L\hat{W}(t)) \qquad (5)$$

$$u_{ad}(t) = -\hat{W}(t)^T \Phi(\eta(t)) \qquad (6)$$

where $\hat{W}(t) \in \mathbb{R}^{n_{nn} \times m}$ contains the weights of the neural network, adapted by the learning algorithm in Eq. (5). L is the learning rate, Φ the vector of basis functions, γ is a forgetting factor, and $e_{nn} = Ke_y$, with K specified later. The projection operator $Proj(p, y(p))$ — which was introduced in Ref. 6 and is now routinely used within adaptive control laws — is a smooth transformation of a given vector field $y(p)$, which gradually adds inward-pointing components to $y(p)$ as the variable p approaches — and surpasses — the limits of a pre-defined compact set.

2.5. *The Adaptive Augmentation*

The "adaptive augmentation" problem can be defined as the problem of using an adaptive element, such as a neural network, in order to totally or partially cancel the unwanted effect of the term Δ. The problem can be also cast as retrofitting an existing linear controller with an adaptive element so that the whole system is improved in terms of performance/stability or robustness.

The adaptive element is tasked to "learn" (and compensate for) the unknown function Δ. If that happens, the forcing terms in Eq. (2) disappear, allowing both $z(t)$ and $e(t)$ to approach zero in absence of the unmatched uncertainties vectors Δ_s and Δ_o. In turn, this means that the overall system is acting as desired despite the presence of the unknown element Δ. Whenever the network cannot cancel exactly the matched uncertainty Δ or either vectors Δ_s or Δ_o are non-zero (but bounded and continuous), the error e(t) will not approach zero but it will however be uniformly bounded inside a compact set.

Formally, the vector $\hat{W}(t)$ containing the neural network weights has to converge to a value $W^* \in \Pi_w$ such that:

$$\Delta = \Delta'(\eta) + \varepsilon_1(\eta) = W^{*T}\Phi(\eta) + \varepsilon_2(\eta) + \varepsilon_1(\eta) = W^{*T}\Phi(\eta) + \varepsilon_3(\eta), \quad (7)$$

where the reconstruction error ε_3 is "as small as possible", and Π_w is the set of "allowed" weights of the neural network. Appropriately defining W_e then yields

$$\Delta + u_{ad} = (W^* - \hat{W})^T\Phi(\eta) + \varepsilon_3(\eta) = W_e^T\Phi(\eta) + \varepsilon_3(\eta) \quad (8)$$

Substituting Eq. (8) in the previous expressions yields the following error dynamics, which includes the neural network contribution:

$$\begin{aligned}\dot{z} &= A_z z + B_u(W_e^T\Phi(\eta) + \varepsilon_3(\eta)) + \Delta_s \\ e_y &= C_z z + D_u(W_e^T\Phi(\eta) + \varepsilon_3(\eta)) + \Delta_o\end{aligned} \quad (9)$$

Therefore, the evolution of the overall system comprehensive of the adaptive element will be completely determined by the two state vectors z and W_e, whose dynamics is specified by Eq. (9) and (5).

3. The Lyapunov Analysis

A system, designed to perform in a steady state, must be stable at times when sudden disturbances affect its operation. Lyapunov analysis is a well known offline approach to evaluating whether the system is able to maintain stability in presence of disturbances. The following general candidate Lyapunov function will be used, where r is a positive scalar, P is positive definite, and tr is the trace operator:

$$V(z, W_e) = \frac{1}{2}z^T P z + \frac{1}{2}tr(W_e^T r L^{-1} W_e) \quad (10)$$

Note that since V is continuous, radially unbounded, and positive, it can be shown that if its time derivative along the system's trajectories is negative outside a certain convex compact set, then the system is *uniformly ultimately bounded* (UUB) within that compact set, and any trajectory approaches the set within a finite time. Such set is therefore indicated as the "UUB set" or "bounding set" of the closed loop system. Calculating the derivative of V with respect to time yields a rather complex quadratic function of the variables z, and W_e. Boundedness can then be proved by showing that this function is negative outside a compact set, and the interplay between the design parameters (such as γ or Q) and the size, dimension and existence of such compact set can then be analyzed in detail.

3.1. *Typical "Completion of Squares" Bounds Formulation and its Limitations*

Several approximations are typically introduced to simplify the involved calculations. Specifically, the quadratic expressions involving the vectors z and W_e are overestimated using the norms $\|z\|$ and $\|W_e\|$, leading to the following inequality:

$$\dot{V}(z, W_e) \leq H(\|z\|, \|W_e\|) \tag{11}$$

where the right hand side H is a quadratic function in the two scalar variables $\|z\|$ and $\|W_e\|$. A conceptual representation of both the surface H and the ellipse $H = 0$ in the two dimensions $\|z\|$ and $\|W_e\|$, is shown in Fig. 2. Note that Eq. (11) implies that the derivative of the Lyapunov function is negative *outside* the ellipse $H = 0$.

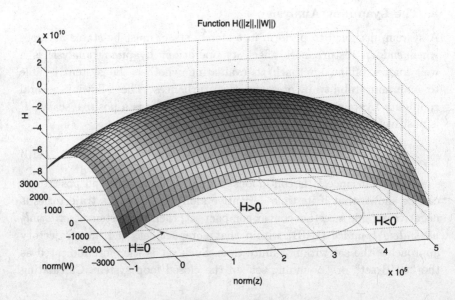

Fig. 2.   Conceptual representation of the quadratic function H.

Additional overestimations are typically carried out at this point, using the fact that for any nonnegative couple of scalars α and β, $\alpha\beta \leq 2\alpha\beta \leq \alpha^2 + \beta^2$, (since $(\alpha - \beta)^2 \geq 0$). Specifically, the terms in Eq. (11) that present a linear- or mixed-dependence on the variables $\|z\|$ and $\|W_e\|$ can be overestimated by three other terms, the first one being a positive constant, the second one featuring $\|z\|^2$ and the third one featuring $\|W_e\|^2$.

Grouping similar terms leads to a quadratic equation having the form

$$H'(\|z\|, \|W_e\|) = a'\|z\|^2 + b'\|W_e\|^2 + g' \tag{12}$$

Whenever a' and b' are negative, the equation $H' = 0$ describes an ellipsoid that is *centered at the origin* and contains the ellipsoid described by $H = 0$, because by construction:

$$\dot{V}(z, W_e) \leq H(\|z\|, \|W_e\|) \leq H'(\|z\|, \|W_e\|) \tag{13}$$

The return set R_E is then defined as the set of points in the space $[z, W_e]$ such that $H' \geq 0$:

$$R_E = \{z \in \mathbb{R}^{n+n_c}, W_e \in \Pi_w | H'(\|z\|, \|W_e\|) \geq 0\} \tag{14}$$

Taking into account Eq. (13), the derivative of the Lyapunov function is negative outside R_E. While this guarantees that any trajectories originating in R_E will eventually return in R_E, it does not guarantee that the trajectories will be fully contained in R_E. It is however guaranteed that the evolution of the system will be confined inside the ellipsoid representing the smallest level surface that fully contains R_E. The bounding set can then be formally defined as:

$$B_E = \left\{z \in \mathbb{R}^{n+n_c}, W_e \in \Pi_w | V(z, W_e) < \max_{\{z, W_e\} \in R_E} V(z, W_e)\right\} \tag{15}$$

Therefore if the initial condition is outside B_E, the overall state $[\|z\|, \|W_e\|]$ asymptotically tends to B_E, while if the initial condition is already within the ellipse, then the system's trajectory remains within B_E. Ideally, the set B_E should be small enough so that $\|z\|$ — which represent the deviation between the desired and the actual states of the system — will remain small, implying that the system behaves as desired. A small B_E guarantees that the behavior of the system from Eq. (3) is confined near the origin, which represents the desired performance. In any case, a constructive method to calculate B_E and visualize its projection along chosen dimensions is crucial, since this allows the requirements and control engineers to quickly check, in the design phase, whether the bounding set intersects sections of the error space — far from the origin — where the behavior of the system is not guaranteed to be safe. In other words, explicit and verifiable requirements could be set, before the design phase, on metrics associated with B_E, for example its size or the size of its projections about selected axis of the error space.

Unfortunately, while Eq. (13) is useful to prove a theoretical point — that there exists a configuration of parameters for which the error is indeed bounded — it is clear that the overestimation of the UUB set obtained in this way is unnecessarily conservative. Furthermore, it has been shown that such estimation can easily lose significance in very common cases — and for sensible selections of the design parameters — for which a bounding set indeed exists. Recent results[21] have shown that H' is not defined whenever the forgetting factor γ is less than 1 and the minimum eigenvalue of Q (defined later) is less than 2, therefore including a majority of practical cases.

For these reasons, current methods to calculate the bounds are not well suited to assess the boundedness of a system in the design phase.

3.2. *A Better Characterization of the Return Set*

By using the following definitions:

$$a = -\frac{1}{2}\lambda(Q)$$
$$b = -r(\gamma - n_{nn}\|KD_u\|)$$
$$c = \frac{1}{2}(\|P\|\|\Delta_s\| + \|PB_u\|\|\varepsilon_3(\eta)\| + 2n_{nn}w_{\max}\sqrt{m}\|M\|) \qquad (16)$$
$$d = \frac{1}{2}r\sqrt{n_{nn}}(\gamma w_{\max}\sqrt{m} + (\|K\|\|\Delta_o\| + \|KD_u\|\|\varepsilon_3(\eta)\|))$$

where $K = \frac{1}{2}B_u^T P C_z^+$ and $A_z^T P + P A_z + Q = 0$, the term H in the inequality Eq. (13) can be conveniently expressed as:

$$H(\|z\|, \|W_e\|) = \begin{bmatrix} \|z\| & \|W_e\| & 1 \end{bmatrix} \begin{bmatrix} a & 0 & c \\ 0 & b & d \\ c & d & 0 \end{bmatrix} \begin{bmatrix} \|z\| \\ \|W_e\| \\ 1 \end{bmatrix} \geq 0 \qquad (17)$$

Specifically, if $a < 0$, $b < 0$, Eq. (16) describes an ellipse such that $H(\|z\|, \|W_e\|) < 0$ for every point that falls *outside* the ellipse. The bounding set B_E is then defined as in (14) and (15) but using directly H instead of H'.

It should be noticed that while it is convenient to think about Eq. (17) as a curve defined in $\mathbb{R}^2$, it is only its intersection with the positive semi-infinite stripe where $\|z\| \in [0, \infty)$ and $\|W_e\| \in [0, 2w_{\max}\sqrt{n_{nn}m}]$ that is relevant, since both variables are only definite within their respective validity intervals.

The Matlab Ellipsoidal Toolbox[22] has a collection of functions useful for the analysis, projection, and visualization of multidimensional quadric surfaces. However, in order to use the toolbox, the set $H(\|z\|, \|W_e\|) \geq 0$ must be expressed as:

$$\Gamma(\xi, q_E, Q_E) = \{\xi \in \mathbb{R}^{(n \,|\, n_v+1)} | (\xi - q_E)^T Q_E^{-1} (\xi \quad q_E) \leq 1\} \tag{18}$$

where $\xi = \begin{bmatrix} \|z\| & \|W_e\| \end{bmatrix}^T$, Q_E is the shape matrix — positive definite for the set in Eq. (18) to be an ellipsoid — and q_E is the vector containing the coordinates of the center. Finally, it is possible to write Eq. (17) in a form suited to using the Ellipsoidal Toolbox, in this case, $\xi = \begin{bmatrix} \|\dot{z}\| & \|W_e\| \end{bmatrix}^T \in \mathbb{R}^2$, $q_E \in \mathbb{R}^2$, and $Q_E \in \mathbb{R}^{2 \times 2}$.

3.3. *Boundedness Conditions*

In this section we present the conditions under which the compact set B_E is closed. A sufficient condition for B_E to be closed is that the set $H(\|z\|, \|W_e\|) \geq 0$ represents an ellipsoid. Using the ellipsoidal calculus formulation, an ellipsoid is defined by the vector q_E and the matrix Q_E where Q_E is positive definite, while a positive semi-definite Q_E characterizes a degenerate ellipsoid. Expressing this condition using the terms defined in Eq. (15) yields the two conditions. The first condition is that the value of a has to be negative. Since Q is definite positive this condition is always verified. The second condition, $b < 0$, guarantees that the ellipsoid does not become degenerate along the $\|W_e\|$ direction.

Whenever the projection algorithm is applied, the B_E set is limited along the $\|W_e\|$ direction to the segment $[0, 2w_{\max}\sqrt{n_{nn}m}]$ because each weight is limited to the interval $[-w_{\max}, w_{\max}]$. However, whenever either no projection algorithm is applied or whenever the limit $w_{\max}$ is selected to be very large, then the condition $b < 0$ is the only mechanism for guaranteeing that B_E is limited along the $\|W_e\|$ direction and hence closed.

Following the definition of b the second condition translates to:

$$\gamma > n_{nn}\|KD_u\| \tag{19}$$

which is — as expected — intrinsically related to the neural network. This condition implies that the forgetting factor must be larger than a certain constant depending on the number of neurons, the matrix K, and the matrix D_u. While this condition is always verified in absence of a matrix D in the original plant, it becomes harder to satisfy whenever the plant has a non-zero D matrix, especially because either higher values of the forgetting factor, or a small number of neurons, may prevent an effective learning process by the neural network. This implies that the matrix K has to be selected small enough so Eq. (19) can be verified. Note, however, that since K multiplies the error seen from the network, ($e_{nn} = Ke_y$), a small K still reduces the effectiveness of the neural network when the other network parameters are kept constant. This confirms that whenever the plant has a non-zero D matrix and the projection operator is not used, Condition (19) implies that the network "freedom" must be kept in check by either enhancing γ or by reducing n_{nn}, or K.

However, it is perhaps more important to consider that, without the application of the projection operator, even when Condition (19) is satisfied no upper bound for $\|W^*\|$ can be selected, therefore leaving the terms depending on $\|W^*\|$ and $\|\varepsilon_3(\eta)\|$ in the ellipsoid expression — specifically the terms c and d in Eq. (15) — undetermined, resulting in an ellipsoid of undetermined size. In this sense, a considerable number of previous boundedness results,[1,17,20] lack constructiveness.

3.3.1. *Extreme Points of the Boundary and Semi-Axes of the Ellipsoid*

The definition of ellipsoid formulated in Eq. (18) provides a useful tool in order to calculate the ellipsoid semi-axes as well as the coordinates of the extreme points of the ellipsoid projection along each axis.

Specifically, the lengths of the semi-axes are the square roots of the eigenvalues of the matrix Q_E, that is:

$$E_j = \sqrt{\lambda_j(Q_E)} \qquad (20)$$

is the length of the jth semi-axis of the ellipse. It should be noticed that the existence of E_j is guaranteed since Q_E is semi-positive definite when the ellipsoids (17) exist.

The coordinates of the extreme points of the projection along the i axis can instead be expressed as:

$$\max(\xi_i) = q_E(i) + \sqrt{Q_E(i,i)}$$
$$\min(\xi_i) = q_E(i) - \sqrt{Q_E(i,i)} \qquad (21)$$

where ξ_i represents the ith component of the vector ξ as defined in previous section, $q_E(i)$ is the ith component of the vector q_E, and, $Q_E(i,i)$ is the component on the diagonal at the position (i,i) of the matrix Q_E. This is in turn of key importance to allow for checking that the UUB set not intersect regions where the behavior is not guaranteed to be safe.

4. Case Study

A simulation study was conducted featuring an adaptive control system for a linear model of an F/A-18 aircraft:

$$\dot{x} = Ax + Bu$$
$$y = Cx + Du \tag{22}$$

The state is described by the following vector:

$$x = \begin{bmatrix} q & \alpha & V & p & r & \beta \end{bmatrix} \tag{23}$$

where V is the airspeed, α and β are respectively the longitudinal and lateral aerodynamic flow angles; p, q, r are the components of the angular velocity in the body reference frame.

The input vector is defined as:

$$u = \begin{bmatrix} ail_l & ail_r & stab_l & stab_r & rud_l & rud_r & tef_l & tef_r & lef_l & lef_r \end{bmatrix} \tag{24}$$

where *ail, stab, rud, tef* and *lef* correspond respectively to ailerons, stabilators, rudders, trailing edge flaps and leading edge flaps, while the l and r subscripts correspond to the left and right actuator.

The output vector is:

$$y = \begin{bmatrix} ancg & axcg & ancs & aycg & aycs & q & \alpha & V & p & r & \beta \end{bmatrix} \tag{25}$$

where *ancg, axcg, aycg* represent the acceleration of the center of gravity respectively along the normal (opposite to z), x and y body axes, and *ancs* and *aycs* represent the acceleration of the center of pressure along the normal and y axes respectively. The other outputs are the state variables.

A non-linearity equal to $-2^* \cos(100 D^T Cx)$ was added to the input to represent a bounded matched uncertainty, while the term $\max(\min(0.1 A^T A_x, l), -l)$ was added to the matrix A to represent a conic-bounded partially matched uncertainty, and where l is the vector $[0.23 \ \ 0.038 \ \ 0.02 \ \ 5.2 \ \ 0.73 \ \ 0.03]^T$. Theoretically, an uncertainty like $\max(\min(0.1 A^T A_x, l), -l)$ can be matched or unmatched, depending on the linear span of the matrix B. In this particular case the matrix B is full row rank and consequently — as it will be explained in more detail in next section — the uncertainty is completely matched. Adding this term causes the system to respond slightly faster to the input commands compared to the reference model. The plant inclusive of the uncertainty is:

$$\dot{x} = Ax + Bu + \max(\min(0.1A^T A_x, l), -l) - 2B\cos(100D^T Cx)$$
$$y = Cx + Du \tag{26}$$

The 10-inputs 6-states 11-outputs plant is controlled by an LQ-servo controller[16], which was designed so that the variables p, q, and r could follow a step input with zero error. The closed loop system involving plant and linear controller, is represented in the orange (upper) block in Fig. 3, while the *reference* closed loop system is instead represented in the green (lower-right) block, and the blue (lower-left) block includes the adaptive element.

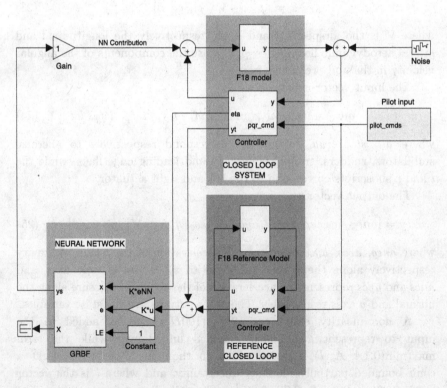

Fig. 3.   Simulink scheme of the overall system.

The Neural Network features 80 standard Gaussian Radial Basis Functions for each of the 10 output channels, with centers chosen randomly between $[-1, 1]$ since the variables in $\eta(t)$ were normalized. The standard deviation of the RBF functions was set to 1, and the matrix W_0 containing the initial conditions of the network weights was set to an initial value derived from a previous off-line learning phase of the Neural Network.

The learning rate L was set to 0.01, the forgetting factor γ was set to 10^{-5}, each element of the weight matrix was limited by selecting $w_{max} = 5$, and the vector $\eta(t)$ includes the 6 states variables of the plant. The error input of the neural network $e_{nn}(t)$ is the error between the real and the reference systems multiplied by K, the constant r was set to 1, and P results from the solution of $A_z^T P + P A_z + Q = 0$ with Q being the identity matrix multiplied by 10^{-6}.

Figure 4 shows the behavior of the real system (dashed line) and the reference system (gray solid line) when a step command in the variable p (black thin line) is provided and the neural network is disconnected. It can be seen that the initial transitory of the real system is quite different to the one of the reference system. Specifically, the real system presents an overshoot of 22% as a response to a positive step and an undershoot of 36%

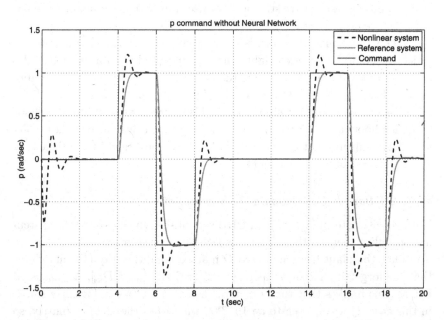

Fig. 4.   Reference and real system responses to a command in p (roll rate).

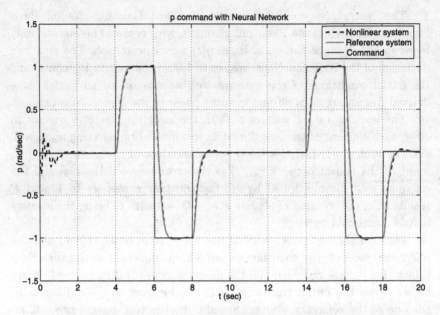

Fig. 5.   Responses to a command in p (roll rate) with Neural Network.

as a response to a negative step, and its rise time is 0.33 sec against the 0.6 sec of the reference one.

When the Neural Network is connected, after an initial transitory in which the Neural Network weights undergo an initial adaptation phase, the behavior of the closed loop system (dashed line in Fig. 5) gets considerably close to the one of the reference model, (gray line in Fig. 5).

It can be seen that the step command is followed without any overshoot or undershoot, and the rise time of the non-linear system becomes 0.654 sec. This confirms that the neural network has converged to a satisfactory approximation of the uncertainty.

4.0.1. 2D *Bounds Calculation and Visualization*

Now we calculate the return and UUB sets of the whole closed loop system and visualize them in the 2D norm space of the system error.

Since the plant has a non-zero D matrix, condition Eq. (19) mandates that the forgetting factor γ has to be larger than $n_{nn}\|KD_u\|$. As discussed before, this implies that the Q matrix has to be selected sufficiently small, in this case, Q was chosen to be $10^{-6}*I$, where I is the identity matrix, so that Eq. (19) could be satisfied, even if, since the projection algorithm is used in the network adaptation laws, satisfying Eq. (19) was not strictly necessary for B_E to be bounded.

In order to calculate the system's bounds in the space of the absolute values of the system error, the parameters a, b, c and d have to be calculated according to Eq. (16). The parameter a is equal to -0.5^*10^{-6}.

An upper bound for the vector of the absolute values of the matched uncertainty Δ can be calculated — in this case — by considering that each element of the vector $-2^* \cos(100 D^T C x)$ cannot be more than 2, and that an upper bound for the absolute value of the other uncertainty element is:

$$| \max(\min(0.1 A^T A_{\dot{x}}, l), \ l)| \leq BB^+ l \tag{27}$$

where BB^+ is the identity matrix, since in this case B is 6 by 11 and has rank 6 (that is B is full row rank), and the values in l were chosen because they are beyond the range of variation of each state variable.

The multiplication of the upper limit by the pseudo-inverse of the matrix B is used to conduct the uncertainty on the state variable x to an uncertainty entering as an input of the system, and therefore totally matched. Of course in applications where B is not full row rank, that is in the majority of the cases, the non-linearity would be mapped as a sum between a matched input uncertainty Δ and an unmatched state uncertainty Δ_s, so that an estimation of $\|\Delta_s\|$ would be necessary.

In any case, the selection of the maximum allowed uncertainty is in general a design choice that is very specific to the application and to the considered control problem.

An upper bound for the elements of the matched uncertainty Δ can be obtained as follows:

$$\|\Delta\| \leq \left\| \begin{bmatrix} 2 \\ 2 \\ \vdots \\ 2 \end{bmatrix} + |B^+|l \right\| = 27.11 \tag{28}$$

In this case being the uncertainty completely matched, the vector $|\Delta_s|$ is equal to zero. Consequently, knowing that $\|\Delta_y\| = 0$, it can be stated that:

$$\|\Delta_o\| = \|D_u \Delta\| \tag{29}$$

and an upper bound for the norm of the best weight matrix is given by:

$$\|W^*\| \leq w_{\max} \sqrt{n_{nn}} \sqrt{m} = 5\sqrt{80}\sqrt{10} = 141.42 \tag{30}$$

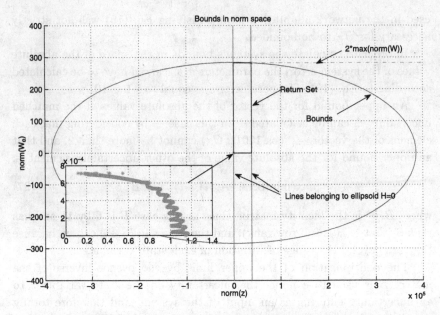

Fig. 6.   Bounds of the system with and without Neural Network in norm space.

Finally the upper bound for $\|\varepsilon_3\|$ can be calculated as:

$$\|\varepsilon_3\| \leq 27.11 \tag{31}$$

Equations (28)–(31) can then be used to calculate the parameters b, c and d in (15), while the value of a has already been calculated.

Figure 6 shows the 2D representation of the bounding set — the greater ellipse — of the overall system comprehensive of the adaptive element. The black rectangle-shaped region contained in the ellipse is the return set, which is the intersection between the semi infinite stripe due to the projection algorithm and the ellipse calculated according to (16) and (17). Note that the curvature of such ellipse is undetectable at the given scale. The green star near the origin indicates the evolution of the system in the norm space, which is expanded in the subplot. The bound of the error $\|z\|$ without the Neural Network can be calculated by considering that the matched and unmatched uncertainties are left uncompensated for, yielding:

$$z_b = \frac{\|P\|\|\Delta_s\| + \|PB_u\|\|\Delta\|}{\frac{1}{2}\underline{\lambda}(Q)} = \frac{2\|PB_u\|\|\Delta\|}{\underline{\lambda}(Q)} = 2.48 \cdot 10^3 \tag{32}$$

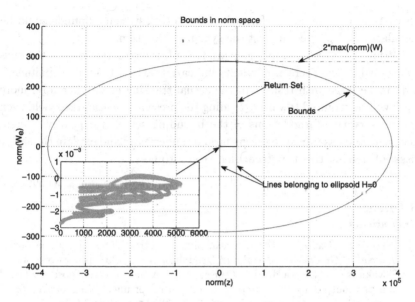

Fig. 7. Bounds and system evolution during malfunctioning.

Finally, a simulation was performed in which the Neural Network weights were appositely set to a vector of random values between -5 and 5, and the learning rate matrix L was set to the identity. As expected, this causes the system to malfunction; the system is actually unstable in the sense that no convergence seems to take place. Furthermore, the value of $\|z\|$ exceeds z_b calculated in (32) indicating that a malfunctioning network could in theory "cause more damage" than the uncompensated uncertainties. However, as shown in Fig. 7, the evolution of the norm of the system error is completely contained within the calculated bounding set.

5. Conclusions

In this chapter, the problem of formulating explicit and verifiable requirements on the evolution of a dynamical system subject to direct adaptive control has been addressed from a control theoretic perspective. Specifically, using methods based on Lyapunov analysis, we developed a general framework to calculate — in the design phase — the bounding set into which the state space of the system is confined at runtime.

The conditions guaranteeing the existence of the bounding set were discussed, and a simulation experiment featuring the adaptive control for an F/A-18 aircraft was performed so the actual system's bounds could be calculated and plotted against the system's evolution.

While the presented methods are much less conservative than the ones considered in the literature, results indicate that the bounding limits still tend to be large with respect to the actual tracking error, therefore, more work will be required toward extending the presented methods to yield even less conservative formulations of the bounding set. Consequently, system designers and assurance engineers need to understand and explain the safety risks inherent in this new technology to customers.

References
Periodicals

1. Kim, B.S., Calise, A.J.: Nonlinear flight control using neural networks, *Journal of Guidance, Control and Dynamics*, 20(1), 1997, pp. 26–33, 144.
2. Hovakimyan, N., Nardi, F., Calise, A.J.: A novel error observer-based adaptive output feedback approach for control of uncertain systems, *IEEE Transactions on Automatic Control*, 47(8), 2002, pp. 1310–1314.
3. Pomet, J.B., Praly, L.: Adaptive nonlinear regulation: Estimation from the Lyapunov equation, *IEEE Transaction on Automatic Control*, 37(6), 1992, pp. 729–740.
4. Hornik, K., Stinchombe, M., White, H.: Multilayer feedforward networks are universal approximators, *Neural Networks*, 2, 1989, pp. 359–366.
5. Park, J., Sandberg, I.W.: Universal approximation using radial-basis-function networks, *Neural Computation*, 3, 1991, pp. 246–257.
6. Campa, G., Fravolini, M.L., Napolitano, M.R.: A Library of adaptive neural networks for control purposes, *IEEE International Symposium on Computer Aided Control System Design*, September 2002, 18–20.
7. Pomet, J.B., Praly, L.: Adaptive nonlinear regulation: Estimation from the Lyapunov equation, *IEEE Transaction on Automatic Control*, 37(6), 1992, pp. 729–740.

Books

8. Palm, W.: *System Dynamics*, McGraw-Hill, September 2004.
9. Esfandiari, R., Vu, H.V.: *Dynamic Systems: Modeling and Analysis*, McGraw-Hill, 1997.
10. Bequette, B.W.: *Process Control: Modeling, Design and Simulation*, Prentice Hall, Upper Saddle River, NJ, 2003.
11. Herbert, R.D.: *Observers and Macroeconomic Systems*, Springer-Verlag, NY, 1998.
12. D'Alessandro, D.: *Introduction to Quantum Control and Dynamics*, CRC Press, Boca Raton, FL, 2007.

13. Narendra, K.S., Annaswamy, A.M.: *Stable Adaptive Control*, Prentice Hall, 1989.
14. Macejowki, J.M.: *Multivariable Feedback Design*, Addison-Wesley, 1989.
15. Jordan, C.: *Calculus of Finite Differences*, Chelsea Publication Company, New York, N.Y., 1960.
16. Stengel, R.F.: *Optimal Control and Estimation*, Dover Publication Inc. New York, 1994.

Proceedings

17. Chang, H., Astolfi, A.: Control of HIV Infection Dynamics, *IEEE Control Systems Magazine*, 28, 2008, 28–39,
18. Rinaldi, S., Gragnani, A.: Destabilizing factors in slow-fast systems, *Ecological Modelling*, 180(4), 2004, pp. 445–460.
19. Campa, G., Sharma, M., Calise, A.J., Innocenti, M.: Neural network augmentation of linear controllers with application to underwater vehicles, American Control Conference 2000, June 2–4, 2000, Chicago, IL.
20. Sharma, M., Calise, A.J.: Neural network augmentation of existing linear controllers, AIAA Guidance, Navigation, and Control Conference and Exhibit, Montreal, Canada, Aug. 6–9, 2001.
21. Campa, G., Mammarella, M., Cukic, B., Gu, Y., Napolitano, M.R., Fuller, E.: Calculation of Bounding Sets for Neural Network Based Adaptive Control Systems, AIAA Guidance Navigation and Control Conference, August 2008, Honolulu, USA.

Computer Software

22. Kurzhanskiy, A.A., Varaiya, P.: Ellipsoidal Toolbox, 2006–2007.

Chapter 7

QUANTITATIVE MODELING FOR INCREMENTAL SOFTWARE PROCESS CONTROL

SCOTT D. MILLER*, RAYMOND A. DECARLO[†] and
ADITYA P. MATHUR*

*Department of Computer Science,
Purdue University
millersc@purdue.edu; apm@cs.purdue.edu

[†]Department of Electrical and Computer Engineering,
Purdue University
decarlo@ecn.purdue.edu

A software development process modeling framework is motivated and constructed under the formalism of State Modeling. The approach interconnects instances of general development-activity modeling components into a composite system that represents the software development process of the target organization. The composite system is then constrained according to the interdependencies in the work to be completed. Simulation results are presented, and implications for control-theoretic decision support (i.e., state-variable control) are briefly discussed.

1. Introduction

The study of software processes has evolved over the decades from prescriptive and descriptive modes into recent applications of predictive modeling/simulation[29] and decision support.[19,17] Techniques range from control charts[30] and statistical process control,[10] where the predictive component is of the nature, "the process should proceed like similar past projects", to system-dynamics models[2,18] and "what-if" scenario evaluation e.g., Ref. 21. Recent work[5,9] has blended techniques from the engineering discipline of control theory into the mix, yielding a contribution to *software cybernetics*;[4,8] it is this work, and particularly Ref. 9, upon which the present work builds.

Scacchi partitions the historical development of *descriptive* and *prescriptive* software process models by whether they detail an actual process that has been executed in the past, or whether they describe characteristics to be implemented in future software development processes. The area of *system dynamics* modeling attempts to capture the causal relations within a process, and to infer from those relations the evolution of the process over time. In this sense, system dynamics modeling layers causal rules over a descriptive model. The system dynamics approach therefore implies a simulation capability wherein the causal rules are automatically applied to transform some representation of an initial process state into an expected future state, proceeding transitively. Where the causal rules are parameterized, one calibrates to match historical data, and extrapolates into the future. The calibration process is called *parameter identification*, which typically takes the form of a weighted error-minimization in shape-fitting. (System dynamics purists may note that system dynamics modeling is also useful for understanding the hidden dynamics captured within the historical data, through parameter identification and *interpolation*).

Where successive parameter re-identification is carried out during the execution of a software development process, the practice is called *quantitative evaluation*. Where quantitative evaluation guides the generation of prescriptive model changes, we identify the practice as *quantitative process control*. Under quantitative process control, schemes which automatically generate prescriptive change suggestions based on a combination of stated objectives, quantitative process state estimation, and predictive model dynamics approach the engineering practice of model-predictive control.[6] It is the explicit goal of the work presented herein to establish a representation of the incremental software development process amenable to the application of model predictive control as applied in control theory.

Specifically, the authors wish to establish a mechanism for decision support within software process control that trades the common stochastic model evaluation methods for analytic ones — that is, rather than aggregating many simulation traces from a stochastic model to determine whether a candidate process change achieves a desired process improvement goal, we seek to derive the process change that best achieves a stated process improvement goal according to a cost functional (or, in engineering parlance, a *performance index*). We have successfully applied such a model predictive control approach[24,26] to the software system-test phase model proposed in Ref. 7; below, we complete the construction of a general model describing incremental software development processes, as begun in Refs. 22, 25, 27.

When building predictive models, care must be taken in selecting the balance between the level of detail captured by the model, the ability to calibrate the model, and the ability to analyze the model. Bounding the spectrum of predictive modeling and simulation techniques are the COCOMO family of models[3] and the class of in-house, organization-specific system dynamics models developed using commercial tools such as Extend.[a] In the case of COCOMO, the prediction is a single scalar project effort estimate. Some detail may be added back into the prediction by use of statistical/heuristic tables that divide the effort estimate into effort per phase, but analysis of the impact of a parameter change in the COCOMO model is still relatively simple given its closed algebraic form. On the other end of the spectrum, custom system dynamics models offer the temptation to capture fine-grained process detail at the cost of large data needs (e.g., to facilitate the calibration of many probability distributions) and near-opacity to analysis.

In the modeling approach taken below, we seek to capture only the dominant dynamics of the software development process under an incremental lifecycle model. We ignore many of the small expected interactions in order to preserve the ability to calibrate and analyze the model; we therefore expect to perform periodic recalibration during the execution of the software development process in order to prevent the accumulation of error in the model state, and thereby, the model predictions.

Broadly, the present work falls under the category of system dynamics modeling. The state of a software development process is represented in a manner consistent with the theory of state variables, and the system dynamics are specified in the form of state-evolution equations. Due to the wide variation in software development processes across industry, the present work describes a tailorable framework for model construction, rather than a single model instance.

---

[a]Extend is a registered trademark of Imagine That, Inc.

1.1. *Contributions*

This work presents the first complete modeling framework for capturing the constraints and behaviors of incrementally scheduled software development through a continuous-time state-model approach — i.e., a step toward applying control theory to the domain of software process control by formulating the software process control problem within the mathematical framework upon which control theoretic techniques have been well studied.

It has been noted[20] that hybrid models are attractive for software process modeling because they naturally model scheduling and dependency satisfaction (i.e., queuing) through discrete-event components, along side the continuous modeling elements which are more natural for phenomena such as productivity. The approach we present captures the scheduling and dependency management within a purely continuous model (as coupled productivity, scheduling, and accumulation components), yielding a simpler simulation and analysis process, and thereby facilitating comprehension and control.

1.2. *Related Work*

With half a century of literature on software engineering, modeling, and control in all their forms, there are too many approaches to cover comprehensively. Two approaches in particular have targeted the goal of control-theoretic decision support based on state models of aspects of software development. Madnick and Abdel-Hamid[2] are credited with an early attempt at applying state modeling techniques for modeling software processes. Their work predicts process behavior from a much higher level (i.e., the behavior of the waterfall development phases) than both the present work, and that described next.

Cangussu[7] proposes a state-based model of the software System Test Phase and supplies a quantitative control technique to aid the manager of a test process in mitigating schedule slippage. The model has been industrially validated, and the controller has been demonstrated through simulation. Indeed, the work by Cangussu *et al.* is considered by the authors to be the direct predecessor to the current work.

For generally related work, the authors direct the reader to the voluminous work in system dynamics modeling, discrete event modeling, and statistical process control as applied to software development processes. Also relevant is the body of literature in the engineering disciplines covering the state-variable modeling approach, state-variable control, model predictive control, and general mixed logical-dynamical system modeling.[16]

2. General Modeling Strategy

Software development can be viewed as a set of seven core development activities occasionally augmented with organization-specific activities. Each activity consumes "work items" of one type in order to produce work items of another type. Such consumer-producer relationships yield the notion of "workflow", as illustrated by numbered arrows in the cyclic directed graph in Fig. 1.

The development activities illustrated in Fig. 1 are defined as follows:

Feature Coding — the creation of product source code from feature specifications.

Test Case Coding — the creation of executable test cases (i.e., automated tests, or scripted instructions) specified within a test plan.

New Test Case Execution — the execution of newly created test cases to exercise the target features, and generation of failure reports as necessary.

Regression Test Case Execution — the execution of previously executed test cases to exercise previously completed product features, and generation of failure reports as necessary.

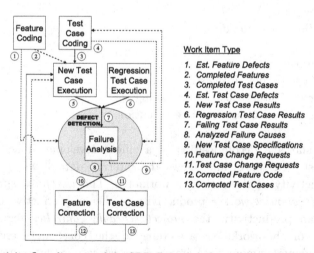

Work Item Type

1. *Est. Feature Defects*
2. *Completed Features*
3. *Completed Test Cases*
4. *Est. Test Case Defects*
5. *New Test Case Results*
6. *Regression Test Case Results*
7. *Failing Test Case Results*
8. *Analyzed Failure Causes*
9. *New Test Case Specifications*
10. *Feature Change Requests*
11. *Test Case Change Requests*
12. *Corrected Feature Code*
13. *Corrected Test Cases*

Fig. 1.   Activity flow diagram of the SDP. Dotted paths are not explicit in the process definition.

Failure Analysis — the analysis carried out upon test-failure events in order to determine the cause of the test failure, and generation of change-requests as necessary.

Feature Correction — the corrective action taken to remediate a defect identified in the functionality of a feature.

Test Case Correction — the corrective action taken by the test developers to remediate a defect identified in the test case coding.

The *defect detection* process estimates the arrival rate of test failure reports given the rates of test case execution and the predicted number of defects present in the software. Note this is an implicit process rather than an explicit development activity; i.e., it does not have a workforce assigned, nor a set of work items to complete.

As mentioned earlier, each "activity" or component in Fig. 1, produces work items. The following list specifies the work items (as numbered in Fig. 1) that flow between activities:

1. *Estimated Defects Produced* — during feature coding.
2. *Completed Features* — for testing.
3. *Completed Test Cases* — for execution.
4. *Estimated Defects Produced* — during test case coding.
5. *Test Case Results* — from new test cases.
6. *Test Case Results* — from regression test cases.
7. *Failing Test Case Results.*
8. *Failures Analyzed.*
9. *New Test Case Specifications*; i.e., when test strategy deficiencies are identified.
10. *Feature Change Requests.*
11. *Test Case Change Requests.*
12. *Corrected Feature Code* — for testing.
13. *Corrected Test Cases* — for re-execution.

Figure 2 shows a decomposition of a single development activity into three constituent parts and shows the internal data flows that govern the productivity of the activity's workforce. The *workforce effort model* represents *potential* worker productivity — hence productive capability rather than productivity; the *project schedule controller* regulates the capability of the workforce according to the constraints imposed by interdependencies among work items, e.g., test case execution must follow test case coding; the *work item queue* tracks the number of work items

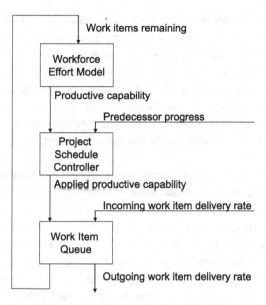

Fig. 2. Modeling a development activity: sub-component data dependencies.

remaining for the development activity, i.e., those work items waiting to be processed.

Note that the project schedule controller receives information about the progress of other development activities. In some cases, it is necessary to prevent an activity from proceeding depending on the status of other activites. For example, test cases may not be executed until the code to be tested has been produced, that is, until the feature coding activity has made a sufficient amount of progress.

Figure 2 illustrates that the workforce members assigned to a given development activity supply a capability for completing work items at a particular rate, this is termed *productive capability*. The project schedule controller then limits the application of this productive capability to work items whose dependencies (*coordination constraints*) have been satisfied (e.g., restricting the productive capability to zero if no work items have their coordination constraints satisfied). Lastly, the *applied productive capability* acts on the pool of work items accumulated in the work item queue to reduce the queue level. Thus one sees that the queue is particularly dynamic in that its level is defined by the balance between the incoming work item arrival rate — supplied by incoming workflows (*in-flows*), and the rate of work item processing dictated by the applied productive capability. The *outgoing work item delivery rate* refers to the rate of by-product generation associated with work item processing, e.g., failure reports are a by-product of test case execution and the defect detection model; such by-product work items are carried away — through *out-flows* — to provide the incoming work item delivery rates to the queues of other development activities in the model. Continuing the example, failure reports flow to the failure analysis task, where their processing generates change requests as a by-product; such change requests then flow to either the feature correction or test correction activity.

Any organization-specific development process can then be represented by an interconnection of these generic activity models of Fig. 2 with coordination constraints imposed to regulate flow rates. A natural framework for this interconnection is the "system-of-systems" modeling approach, shown in Fig. 3. Here, the seven development activity models are illustrated as independent subsystems (labeled S_i for $i = 1, 2, \ldots, 7$); we assume (and later substantiate) that each development activity model can be encapsulated into a continuous time state-model representation characterized by a local input vector (designated a_i) supplying the inputs to each of the three sub-components of the development activity model, a local output vector (designated b_i), and a local state vector (designated x_i). System-of-systems composition allows algebraic subsystems as well; that is, subsystems that possess no internal state. Figure 3 illustrates one such algebraic process: the defect detection model, A_{dd}.

In Fig. 3, the individual input, output, and state vectors for the subsystems (i.e., for the development activity models and defect detection

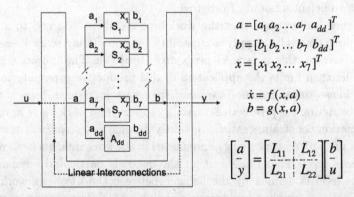

$$a = [a_1\, a_2 \ldots a_7\, a_{dd}]^T$$
$$b = [b_1\, b_2 \ldots b_7\, b_{dd}]^T$$
$$x = [x_1\, x_2 \ldots x_7]^T$$

$$\dot{x} = f(x, a)$$
$$b = g(x, a)$$

$$\begin{bmatrix} a \\ \hline y \end{bmatrix} = \begin{bmatrix} L_{11} & L_{12} \\ \hline L_{21} & L_{22} \end{bmatrix} \begin{bmatrix} b \\ \hline u \end{bmatrix}$$

Fig. 3. System-of-systems assemblage of modeling components with linear interconnections.

model) are concatenated into the vectors a, b, and x, called the *subsystem input vector*, and the *subsystem output vector*, and *subsystem state vector*, respectively. Having a definition for a and b, one may now consider the mechanism for interconnecting the subsystems: The system-of-systems framework illustrated in Fig. 3 explicitly assumes that all interconnections are linear in nature; so-called "nonlinear connections" can be absorbed into an algebraic subsystem, as is the case with algebraic portions of the defect detection model. The matrix L and its four partitions L_{ij} (for $i, j \in \{1, 2\}$) define a linear combination of elements from the subsystem output vector and the *external input vector* (designated u) that generate the elements of the subsystem input vector, and the *external output vector* (designated y); from the definitions one observes that all vectors are time-dependent.

Examining L, one notes that the partition L_{11} describes the linear interconnections from the subsystem outputs to the subsystem inputs; L_{12} describes how the external inputs enter linearly into the subsystem inputs; L_{21} specifies the linear combination that generates the external output from the subsystem outputs; and lastly, L_{22} specifies any linear feed-through of the external input to the external output. Finally, the dynamics of the system can be expressed in terms of a, b, and x in the following vector valued functions:[13,14] The time-derivative of the subsystem state vector is given (using Newton's notation) as a general non-linear function of the subsystem state and the subsystem input vector: $\dot{x} = f(x, a)$; the subsystem output vector is defined by a general non-linear function of the subsystem state vector and the subsystem inputs as $b = g(x, a)$. Thus, if a state model representation of each development activity can be described, and the interconnections between them are linear, then there exists a concise and elegant representation of the overall interconnected system-of-systems.

Under such a framework, it is clear that each modeling component can be described independently and subsequently be assembled into the target system by specifying (i) the linear interconnections (feed-forward and feedback), (ii) the incorporation of external inputs, and (iii) the generation of the external outputs, through the matrix L. In the development that follows, we begin by constructing the sub-components of a generic development activity model (as given in Fig. 2) and then show how the generic activity models are composed into a model of the form of Fig. 3. Following the mathematical model development, we present an example model and simulation results.

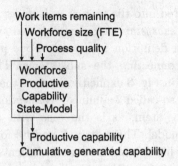

Fig. 4.   I/O description of the workforce productive capability.

2.1. *Mathematical Modeling of Productive Capability*

Figure 4 indicates that the level of productive capability produced by the workforce is dependent on the number of work items remaining, the number of full-time equivalent employees (FTEs) assigned to the development activity, and a process metric, called "quality", that captures the impact of the organization's operating procedures on the rate at which the workforce can process work items.

In the following development, *productive capability* refers to the *potential* for the workforce to complete work items at a given rate. For example, one may hire twenty qualified developers and achieve a high potential productivity — but unless one has work for them to perform, they achieve zero productivity — i.e., the potential over any period of time they sit idle is wasted; this is the key difference between productive capability and productivity.

Equation (1) sets forth a method for computing productive capability

$$\dot{\rho}_t = e^{-(\alpha+\theta_t\beta)^2} F^{cap}\omega_t - \frac{\xi}{\gamma_t}\rho_t \tag{1}$$

where ρ_t is productive capability, ω_t denotes workforce size, θ_t is the remaining number of work items, γ_t is the process quality, and F^{cap} denotes the calibrated average amount of productive capability per FTE. The other variables are data-dependent calibration parameters, addressed later. Note that the workforce is specified in FTEs — the intent is that a subject-matter expert can be counted as the equivalent of more than one general software engineer; this is an explicit solicitation for the use of manager knowledge regarding the capabilities of the specific workforce assigned to an activity.

An intuitive motivation for this equation uses an analogy with classical mechanics, in the style of Ref. 9, as follows: Change within the software development process (SDP) takes place in a continuous manner by the application of a "force" until the desired outcome is reached. Let the term *process inertia* within the SDP refer to the human tendency to resist change — that is, the tendency for employees to settle into a comfortable rhythm. In this analogy, *physical mass* can be equated to *workload complexity*. Let *physical position*, as an offset from the origin, be analogous to the count of work items that a team must complete; then the act of completing a number of work items of a certain complexity is analogous to displacing an object of a certain mass a certain distance toward the origin; velocity becomes the rate of work item completion, and momentum corresponds to the rate of work item completion weighted for complexity of the work items (and hence is a normalized measure of productivity).

A macro-description of productive capability is therefore given as the force-balance equation,

$$F_t^{net} = F_t^{act} - F_t^{res} \tag{2}$$

where F_t^{net} is the net force responsible for acceleration, F_t^{act} is the actual force applied by the workforce, and F_t^{res} is a momentum-dependent resistive force, i.e., the difficulty that the workforce encounters in completing work items of a given complexity at a given rate. It may be useful to consider the resistive force to be the amount of productive capability consumed in communication and keeping up with code changes; then it is clear that if either the complexity of the work items or the rate of their completion increases, then a larger productive capability must be supplied by the workforce if the balance is to be maintained. We model this resistive force as a productivity-dependent damping force opposing the effort provided by the workforce assigned to an activity; hence the dashpot-like damping equation,

$$F_t^{res} = \frac{\xi}{\gamma_t} \rho_t \tag{3}$$

where ρ_t is the instantaneous momentum, ξ is a calibrated measure of the resistance force encountered per unit of momentum — note that the effect of ξ is influenced by a process quality variable, $0 < \gamma_t \leq 1$. Whereas ξ is intended to characterize the difficulty encountered by the specific team working in the application domain of the given development activity,

γ_t captures the impact of aspects of the working environment such as teaming/mentoring, operating procedures, meetings, etc. — those aspects under the control of project management which can distract, interrupt, or otherwise cause the workforce to work at less than optimal efficiency (i.e., $\gamma_t < 1$).

Additional workforce FTEs are assumed to provide a development activity with a proportionally larger capability to do work; in particular, Eq. (2) describes the peak potential force that the workforce can produce

$$F_t^{pot} = F^{cap}\omega_t \tag{4}$$

where F_t^{pot} is the instantaneous potential force supplied by the team; F^{cap} is a calibrated measure of the average force contributed per FTE, and ω_t is the number of FTEs allocated to the development activity.

Unfortunately, workforce members do not always realize their peak performance; Czikszementmihalyi[12] provides the final piece; Czikszementmihalyi proposes that human productivity is affected by both extremes of the "level of challenge". If the challenge level is too high or too low, productivity levels fall; the full capability of the workforce is realized only when the level of challenge is balanced to the capability of the workforce. The Gaussian function provides a ready match to the anecdotal description of the phenomenon, and through careful calibration, a segment of the Gaussian function can be selected to capture the observed (historical) productive capability of a given development activity's workforce as a function of the amount of outstanding work waiting in the queue for the development activity; Fig. 5 illustrates an example segment of the Gaussian curve, and gives the parameters that define its shape. Here we make an explicit assumption that Czikszementmihalyi's notion of "challenge" applies to a development activity's sense of "backlog pressure"; that is, we assume that the challenge of meeting management demands on clearing the backlog of work items yields the same sense of "overwhelming" or "underwhelming" responsibility that drives productivity. Certainly with minima of backlog length, one expects the workforce to feel less challenged in meeting the management deadlines; for maxima, there is certainly room for the same fear of incompleting an extremely complex task; in either case one may expect procrastination and avoidance behaviors. We therefore build a model of "applied force" by reducing the potential force supplied by FTEs down to that portion which is expected to actually be applied to process the work items rather than performing avoidance behaviors. In Eq. (5), a segment of the Gaussian function is selected via the calibration parameters to represent

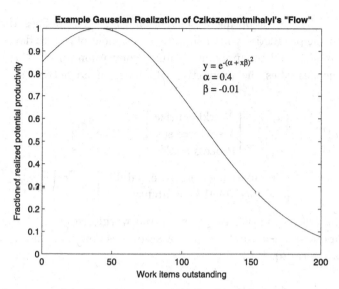

Fig. 5. A segment of the Gaussian curve, shifted and stretched to an x-axis in units of outstanding work items. For the arbitrarily selected example parameters, an outstanding workload of approximately 40 work items provides the maximal expected realization of the workforce potential to do work.

the range of fractions of the workforce's peak capability that one expects to be realized given an expected range in the magnitude of the outstanding workload:

$$F_t^{act} = e^{-(\alpha + \theta_t \beta)^2} F_t^{pot} \qquad (5)$$

where F_t^{act} is the instantaneous *actual* force applied, θ_t enumerates the remaining work items, and α and β are parameters calibrated to historical data — α displaces the center of the curve, β controls the width of the bell shape. That is, the Gaussian curve is shifted and stretched such that, when evaluated at a point corresponding to the outstanding workload (i.e., level of an activity's queue), the curve height sets the fraction of peak productive capability expected to be realized as useful work. Equation (5) is therefore a quantitative approximation of Czikszementmihalyi's challenge-based productivity anecdote.

2.2. State Model of Productive Capability

Noting that instantaneous net physical force applied to an object is equal to the instantaneous derivative of the object's momentum, i.e., $F_t^{net} = \dot{\rho}_t$, one arrives at Eq. (1) via substitution of Eqs. (2)–(5). In

State Model form, adopting a convention that a_i^{prod} defines the input vector for the productivity modeling sub-component of the ith development activity model, and by applying a similar convention for b_i^{prod}, x_i^{prod}, and $\dot{x}_i^{prod}$, one describes the productive capability of an activity's allocated workforce as

$$a_i^{prod} = \begin{bmatrix} a_1 \\ a_2 \\ a_3 \end{bmatrix} = \begin{bmatrix} \text{Workload size} \\ \text{Workforce size} \\ \text{Process quality} \end{bmatrix}; b_i^{prod} = \begin{bmatrix} b_1 \\ b_2 \end{bmatrix}$$

$$= \begin{bmatrix} \text{Cumulative generated capability} \\ \text{Productive capability} \end{bmatrix} = \begin{bmatrix} x_1 \\ x_2 \end{bmatrix} = x_i^{prod}$$

where the state evolution equation for the workforce productivity model sub-component of the ith development activity is now given in Eq. (6) (and is equivalent to Eq. (1).)

$$\dot{x}_i^{prod} = \begin{bmatrix} \dot{x}_1 \\ \dot{x}_2 \end{bmatrix} = \begin{bmatrix} x_2 \\ e^{-(\alpha + a_1\beta)^2} F^{cap} a_2 - \frac{\xi}{a_3} x_2 \end{bmatrix} \tag{6}$$

2.3. *State Model of a Queue*

To track work item accumulation and to facilitate enforcement of the coordination constraints, a non-standard queuing semantics is defined below and illustrated in Fig. 6. The inputs to the queue model are

1. the *in-flow rate* — the rate at which work items arrive from "up-stream" development activities, and

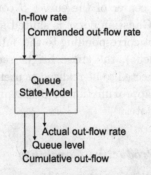

Fig. 6.   I/O description of a work item queue.

2. the *commanded out-flow rate* — the rate of out-flow from the queue commanded by the scheduling controller (i.e., the requested rate of work item processing).

The outputs of the queue model are

1. the *actual out-flow rate* — the rate at which work items are actually leaving the queue to be processed by the workforce (i.e., the actual rate of work item processing),
2. the *queue level* — the number of work items currently in the queue, and
3. the *cumulative out-flow* — the time-integral of the out-flow rate.

In general, the queue level is computed as the integral of the difference between the in-flow rate and the actual out-flow rate. The actual out-flow rate is typically equal to the commanded out-flow rate, so long as the queue is not empty. As the queue level approaches zero, the actual out-flow rate approaches the lesser of the commanded out-flow rate and the in-flow rate. Intuitively, if the commanded rate is non-zero when the queue level is zero, then the queue input is passed directly to the output. In such scenarios, the actual out-flow rate is limited to the in-flow rate under high demand from an empty queue.

In state-model form, the queue for the ith development activity is described as

$$a_i^{queue} = \begin{bmatrix} a_1 \\ a_2 \end{bmatrix} = \begin{bmatrix} \text{In-flow rate} \\ \text{Commanded out-flow rate} \end{bmatrix};$$

$$b_i^{queue} = \begin{bmatrix} b_1 \\ b_2 \\ b_3 \end{bmatrix} = \begin{bmatrix} \text{Cumulative generated out-flow} \\ \text{Queue level} \\ \text{Actual out-flow rate} \end{bmatrix} = \begin{bmatrix} x_1 \\ x_2 \\ x_3 \end{bmatrix} = x_i^{queue}$$

with dynamics

$$\dot{x}_i^{queue} = \begin{bmatrix} \dot{x}_1 \\ \dot{x}_2 \\ \dot{x}_3 \end{bmatrix} = \begin{bmatrix} x_3 \\ a_1 - x_3 \\ \frac{1}{\tau}\left(\text{sm}(a_1, a_2) + \text{sat}(x_2)(a_2 - \text{sm}(a_1, a_2)) - x_3\right) \end{bmatrix} \quad (7)$$

constrained by $x_i \geq 0$, $i \in \{1, 2, 3\}$ and $a_j \geq 0, j \in \{1, 2\}$; where τ is a time-constant, $\text{sm}(\cdot)$ is a smoothed approximation of the $\min(\cdot)$ function built from $\text{sat}(\cdot)$, i.e.,

$$\text{sm}(a, b) = a + \text{sat}(b - a)(b - a) \quad (8)$$

and sat($\cdot$) is a piecewise continuous saturation function,

$$\text{sat}(x) = \begin{cases} 0 & \text{if } x < 0 \\ x & \text{if } 0 \le x \le 1 \\ 1 & \text{if } 1 \le x \end{cases} \tag{9}$$

The state element x_2 tracks the queue level, and hence is constrained to be non-negative. It is computed as the integral of the difference between the in-flow rate, a_1, and the actual out-flow rate of the queue, x_3. Because it is likely that an empty queue will be supplied with a positive commanded out-flow rate, the RHS of the second equation in Eq. (7) cannot simply integrate the difference between the in-flow and commanded out-flow rate, (i.e., $a_1 - a_2$). For this reason, x_3 is introduced as a slack variable to allow the out-flow rate to satisfy the physical constraints of queue behavior.

The third equation in the RHS of Eq. (7) regulates x_3 — and correspondingly, the actual out-flow rate of the queue — so that tracking of the commanded rate, a_2, occurs when possible, i.e., when $\text{sat}(x_2) > 0$. Howeever, on the occasion that the queue is empty, i.e., $\text{sat}(x_2) = 0$, the regulator causes x_3 to track the minimum of the in-flow rate and the commanded out-flow rate in order to preserve the constraints.

2.4. *Normalization of Work Items*

The mapping from work items within the model to the things they represent in the software development process has not yet been rendered explicit. Recall the physical analogy developed in Section 2.1, by which the number and complexity of work items are related to the concepts of distance and mass. Recalling the definition of work from classical mechanics as force times distance, or equivalently, mass times acceleration times distance, we see that for a given acceleration, mass and distance are interchangeable. That is, to accelerate a mass m over a distance d with acceleration a requires an equal amount of work to accelerating a mass $k \times m$ over a distance $\frac{d}{k}$ with the same acceleration a. This observation motivates the normalization of work items to a fixed unit of complexity. That is, a single work item with complexity 5.3 is instead treated as if it were 5.3 work items of unit complexity. Because this conversion does not affect the notion of work within our physical analogy, it has no effect on the model, other than allowing the equations to track momentum (complexity-normalized productivity), rather than a separate velocity (work item completion rate) and mass (complexity).

The resulting notion of unit complexity may aid the modeler in specifying conversion factors for by-product generation (e.g., how many test case execution work items are produced upon the completion of a test case authoring work item?). Such conversion factors appear when specifying the workflows between development activities; for simplicity, and without loss of generality, we present the model construction under the assumption that the processing of one work item generates one by-product work item; in practical applications, conversion factors are applied in the linear interconnection matrix.

2.5. *Managing Dependencies and Scheduling Constraints within the Model*

As mentioned earlier, the execution of a given development activity may be dependent on the progress of another development activity. For example, test execution may be dependent on the progress of feature coding; that is, testing activities may only proceed over the set of test cases that can be meaningfully executed given the state of completion of the features that the tests are intended to exercise. The ability to capture and incorporate such constraints into the model becomes particularly relevant when considering incremental software development processes, in which each development activity may have a sequence of scheduled internal milestones. If circumstances allow the development activities to proceed asynchronously, then one could be modeling a development effort where the developers are coding features for internal release 2, while the testers are running tests on release 1 features, and while the test case coders are writing tests for the third internal release.

In the typical free-form system dynamics modeling approaches, one would simply replicate the development activity models for each release, and switch them on or off by reallocating the workforce to "power" them as deemed appropriate. Such workforce reallocation may be scripted, depending on certain conditions being satisfied by the state of the model, or implemented via manual observation and intervention. In either case, the model becomes significantly larger, and in many cases, more opaque to analysis. Consider, for example the seven development activities described above: Considering only the queue and productivity models, each development activity model has 5 inputs and 5 state elements, for a total of 35 state elements over the 7 processes. If one were to replicate each activity for three internal releases, the system would need to manage 105 inputs and state elements. Because our long term goal is to develop a reasonably accurate model with properties amenable to analysis and parameter identification, we have chosen a different approach.

We have taken the view from the workforce member's perspective: there is not a clear-cut time instant at which the workers stop all release 1 work items and switch solely to release 2 work items; rather, we assume that developers are assigned a set of tasks among which the individual workforce members manage their own time. Under this model, work items from release 2 begin to enter the development process at the same time that release 1 work items are being completed. Indeed, if there were a strict boundary between the two releases, the entire workforce assigned to a development activity would need to sit idle while the last work item from release 1 was being completed by a single developer; it is also unlikely that a fixed fraction of workforce time is walled off and dedicated solely to release 2 tasks once they become available, as would be the interpretation of the scheme using workforce reallocation across replicated development activity models. In avoiding these undesirable assumptions about the target development process, our scheme sacrifices the detail about which specific tasks are being completed. For developers managing their own time among a collection of tasks, it is unlikely that any model can accurately reflect which work items are receiving effort at a given time. We track only the amount of normalized work items completed; the modeler must determine, based on history, practice, operating procedure, etc., the amount of progress that a development activity must achieve to signify that a particular milestone has been reached.

For example, consider a project with two incremental releases, R1 and R2, to be implemented serially. Assume that R1 has 10 features and 70 test cases and R2 has 20 new features and 130 more test cases. Recall the earlier discussion that test execution requires a completed set of features to test; a modeler may specify that once the feature coding development activity has completed the number of work items corresponding to the first 10 features, then the first 70 test cases become available to the test case execution activity. Alternatively, the modeler may opt to make the R2 feature coding work items available slightly sooner to represent a less strict implementation of serial releases. Correspondingly, delays in the transition between internal releases can be modeled by postponing the availability of work items to an activity; e.g., the release of R1 test execution work items can be delayed until feature coding completes a few R2 work items.

We now introduce the notion of *coordination constraints* into the model; intuitively, coordination constraints specify a threshold on the amount of progress (i.e., cumulative out-flow) that a development activity is allowed to make.

2.6. *An Algebraic Model of Activity Coordination*

Coordination constraints are implemented by synthesizing a controller, illustrated in Fig. 7, that causes the *cumulative out-flow* of the dependent queue to honor a user-specified threshold function — essentially placing a regulator on the flow of work items from the dependent queue. For simplicity, the threshold function must satisfy the following conditions to be permissible:

1. The threshold may be constant, or it a may be a function of the cumulative out-flow outputs of other queues in the interconnected system.
2. The threshold must be monotonically increasing with time, and must be defined everywhere in the state space.
3. The threshold must be time-differentiable everywhere in the state space.

The first condition is placed to ensure that the threshold can be evaluated in the presence of interconnections; it is otherwise conceivable, though meaningless, to define the threshold as a non-linear function of its own output. In such a case, zero or perhaps multiple solutions could result, and the task of determining a solution could become arbitrarily hard. By placing the first condition, the threshold is constrained to

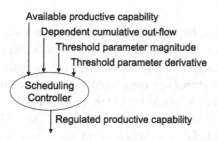

Fig. 7.   I/O description of the project schedule controller.

the intended purpose of coordinating work item interdependencies in an environment of asynchronous progress among development activities. The second condition ensures that invalid states cannot result from the choice of threshold; without such a condition, one may produce a threshold function that, at some point, lowers the threshold below the monotonically increasing cumulative output of the queue — this would yield an invalid state. The third condition arises from a consideration similar to the behavior of an empty queue under a non-zero commanded out-flow rate, as described in the queue model development above: When the progress of a development activity has reached the threshold specified by a coordination constraint, then the *regulated productive capability* (i.e., the amount of productive capability passed from the workforce productivity model through the project schedule controller) is reduced to the lesser of the productive capability generated by the workforce productivity model, and the rate of change (i.e., time-derivative) of the threshold imposed by the coordination constraint. Recall that the threshold may be a function of time-dependent outputs from other development activities, hence one must allow progress commensurate with the rise of the coordination constraint threshold.

Considering the example two-release incremental process described above, one could specify the threshold for the test case execution task as a function of the feature coding progress (assuming 1 feature = 1 work item and 1 test case = 1 work item, for simplicity) as

$$c(x) = \begin{cases} 0 & \text{if } x < 9 \\ -140x^3 + 210x^2 & \text{if } 9 \leq x < 10 \\ 70 & \text{if } 10 \leq x < 29 \\ 260x^3 + 390x^2 + 70 & \text{if } 29 \leq x < 30 \\ 200 & \text{if } x \geq 30 \end{cases}$$

Here, the threshold function, $c(x)$, is given as a step function with the step transitions smoothed by cubic splines. Using such a threshold function, it is trivial to see that the threshold rises to 70 once the cumulative feature coding progress reaches 10 work items; the next transition occurs after the additional 20 R2 features are completed (i.e., a cumulative progress of 30 features), raising the threshold by 130 to a total of 200. The time-derivative of such a threshold function can be computed using the chain rule, i.e., $\frac{d}{dx}c(x)\frac{dx}{dt}$, thus, so long as a measurement of $\frac{dx}{dt}$ is available, computation of the time-derivative to satisfy the third condition is trivial. Consider now that the first condition only allows the threshold to be constructed from constants and interconnections to the "cumulative out-flow" output elements of other queues — the time derivatives of these terms are always available, either as 0 for the constants, or as the "actual out-flow rate" output elements from the same queues.

In synthesizing the controller, there are a few cases to consider: (i) the dependent queue's cumulative output is below the threshold — no regulation is required; or (ii) the controlled queue's cumulative output is at the threshold — if the threshold has a non-zero time derivative, then the controlled queue's actual output rate may take a value between zero and the time-derivative of the threshold. Again, the preceding cases should be reminiscent of the controller embedded into the queuing equations; in fact the controllers are of the same form.

The algebraic control law implementing the scheduling controller for the ith development activity is given by

$$a_i^{ctrl} = \begin{bmatrix} a_1 \\ a_2 \\ a_3 \\ a_4 \end{bmatrix} = \begin{bmatrix} \text{Productive capability} \\ \text{Dependent cumulative out-flow} \\ \text{Threshold parameter magnitude} \\ \text{Threshold parameter derivative} \end{bmatrix}$$

$$b_i^{ctrl} = [b_1] = [\text{Controlled capability}]$$

where

$$b_1 = \text{sm}\left(\left.\frac{d}{dz}c(z)\right|_{a_3}, a_4, a_1\right) + \text{sat}(c(a_3) - a_2)\left(a_1 - \text{sm}\left(\left.\frac{d}{dz}c(z)\right|_{a_3}, a_4, a_1\right)\right)$$

for the user specified threshold function $c(\cdot)$.

2.7. *Defect Modeling and the Failure Analysis Activity*

A careful distinction must be made about the role of the failure analysis activity: In the SDP, failure analysis is the process of determining the source of a test case failure (a feature defect, or a test case defect?) — yet the generic activity model developed thus far is only responsible for estimating the *effort* and/or *duration* encountered in a given activity. To adequately represent the failure analysis activity, the generic development activity model must be augmented with a system for estimating the rate of test case failure, and for sorting failures by their respective causes —

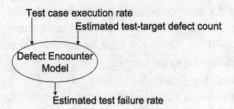

Fig. 8.　Description of the general defect detection model.

i.e., a defect detection model. Such a defect detection model is illustrated in Fig. 1, encapsulating the failure analysis activity.

We define a *defect encounter model* as a non-linear algebraic component with inputs and outputs as shown in Fig. 8. Its general form is adapted from Ref. 7, which builds upon the Lotka–Volterra predator–prey population dynamics model.[31] The analogy proposed in Ref. 7 asserts that the defects present in a software target form a population of prey; the test cases executing against the software give a population of predators. By enumerating the possible encounters and assigning a probability of occurrence, one can construct a model for predicting average defect detection rates (as validated in Ref. 9). Because test execution is temporal, the rate of test execution is analogous to the number of predators hunting in the territory of the prey per unit time; hence defect detection is computed in the present model as an expected rate of encounter between defect/test case pairs. The general form of the algebraic defect detection equation is given in Eq. (10).

$$r = \mu\, e\, d \qquad\qquad (10)$$

where e is the test execution rate, d is the estimated number of defects presently in the test target; hence, their product gives the potential encounters per unit time. The calibrated parameter μ gives the probability that a potential encounter actualizes; that is, μ gives the probability that any given test case will reveal any given defect. This generalization is only applicable when considering the interaction of a collection of tests against a collection of defects, which is the present use. The value r, on the left-hand side, gives the expected test case failure rate. It is clear that every term in Eq. (10) is time-dependent, however it may be sufficient to treat μ as a piecewise constant parameter, adjusted periodically through re-calibration — an approach validated empirically in Ref. 9.

The appearance of the variable d implies a defect estimation component in the model. The estimated number of defects present in the test target may be determined by analysis of defect detection data,[7] predicted from process factors — with their impact calibrated from past projects,[11] or modeled through defect introduction/removal dynamics. Any of these approaches, and indeed hybrids of them, fit well within the system of systems composition framework. Without loss of generality, we shall assume that defect introduction is modeled as per the nonlinear defect introduction component of COQUALMO[11] (details in Section 2.8). Defect detection is modeled per Eq. (10) and defect removal is tracked by the cumulative out-flows from the feature and test correction activities, respectively, with appropriate scaling (per Section 2.4). The difference between the estimated *defects introduced* and the estimated *defects removed* therefore provides an estimate of the number of *defects remaining, d*.

In Fig. 1, two instances of the general defect encounter model are implied: one that encounters defects in the feature code, and another that encounters defects in the test case code. The separation of these cases is important; the ratio of these cases present in the workload of the failure analysis activity precisely defines the ratio of each case evident in the out-flow — that is, this ratio defines how to split workflow 8 (in Fig. 1) into workflows 10 and 11.

To model this split, we construct an algebraic proportional splitting component, illustrated in Fig. 9, satisfying:

$$|\text{Source}| = |O1| + |O2|$$
$$|R1| : |R2| = |O1| : |O2| \tag{11}$$

To capture the distinction between test case failures due to feature defects and test case failures due to test case defects, the input queue of the failure analysis process is split into two queues, each fed by its respective

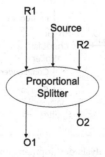

Fig. 9. I/O description of the algebraic splitting component required to separate the out-flows of the failure analysis activity into the in-flows for the correction activities.

defect encounter model. The queue levels can then provide the reference inputs to the splitting component, with the source input supplied as the commanded out-flow rate from the failure analysis schedule controller. Note that a project manager has no visibility into the particular queue from which a given failure analysis work item is drawn; indeed, it is the explicit purpose of the failure analysis activity to make this determination. We therefore assume that the work items in each queue have equal probability of being drawn — that is, the out-flow rate, as regulated by the project schedule controller, is split proportionally — according to the relative queue levels — before being applied to the respective queues. To represent the size of the outstanding workload required by the workforce effort model within the failure analysis activity, the two queue levels are simply added. Figure 10 illustrates the failure analysis activity model described above. Note particularly the summation of the queue levels to provide the number of remaining work items to the workforce effort model, and

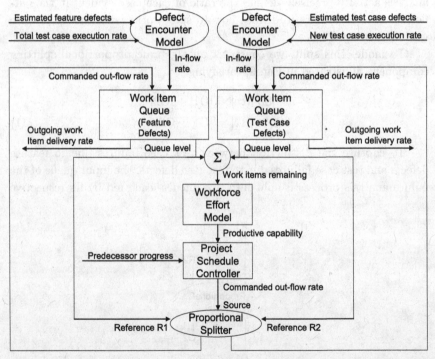

Fig. 10. Modeling the failure analysis development activity: Data dependencies among sub-components.

the embedded usage of the proportional splitting component to divide the commanded out-flow rate between the queues in proportion with their relative queue levels.

2.8. *An Algebraic Model for the Example Defect Population Estimation Component*

The form of the defect introduction sub-model of COQUALMO[11] is given in Eq. (12),

$$\sum_{j=1}^{3} A_j * (Size)^{B_j} * \prod_{i=1}^{21} (DI\_drivers)_{ij} \tag{12}$$

where j iterates over the three defined categories of development artifact types, and i over the impact of 21 different process factors describing the process used to create each type of artifact (called DI_drivers, or *defect introduction* drivers). The DI_drivers affect the amount of defect introduction expected per unit adjusted artifact size. Here, A_j is a scalar calibration constant defined for each artifact type, and B_j is the COCOMO-like constant scale-factor capturing the artifact-type dependent economies or diseconomies of scale with respect to defect introduction. Size in Eq. (12) refers to the size of the artifacts as captured through theoretical metrics (e.g., function points[1]) or through direct measurement (e.g., kSLOC[28]).

In our model, we examine only the coding artifacts, and assume that the impact of the cost drivers and the scale factor have been pre-computed. We therefore simplify Eq. (12) to the form $As^b D$ before including it in the development.

Given inputs defined as

$$a^{dp} = \begin{bmatrix} a_1 \\ a_2 \\ a_3 \\ a_4 \end{bmatrix} = \begin{bmatrix} \text{Features produced} \\ \text{Feature defects removed} \\ \text{Test cases produced} \\ \text{Test case defects removed} \end{bmatrix}$$

The count of defects in the feature code and test case code is therefore modeled as

$$b^{dp} = \begin{bmatrix} b_1 \\ b_2 \end{bmatrix} = \begin{bmatrix} \text{Estimated Feature Defects Present} \\ \text{Estimated Test Case Defects Present} \end{bmatrix}$$

$$= \begin{bmatrix} A_{fc}(a_1)^{b_{fc}} D_{fc} - a_2 \\ A_{tc}(a_3)^{b_{tc}} D_{tc} - a_4 \end{bmatrix}$$

where subscripts of *fc* and *tc* call out calibration parameters from CO-QUALMO, tailored to the feature coding and test case coding artifacts/team/process, respectively.

Analyzing the derivative of b^{dp} with respect to changes in artifact size, we find non-linear growth in the number of estimated defects introduced as a function of total size; this finding is in accord with conventional wisdom in defect modeling. Hence, by driving inputs a_1 and a_3 from the (appropriately scaled) cumulative out-flows of the feature coding and test case coding activities, respectively, one achieves a COQUALMO-compatible defect introduction estimate that evolves over time. Correspondingly, if a_2 and a_4 are driven from the (appropriately scaled) cumulative out-flows of the feature correction and test case correction activities, then one achieves a dynamic estimate of the instantaneous number of defects present in the features and/or test cases.

2.9. *An Algebraic Model of the Defect Detection Component*

The following algebraic defect detection model estimates the arrival rate of test case failures for both categories of cause defined above (i.e., feature defect or test case defect). We define the local input vector of the defect detection component as

$$a^{dd} = \begin{bmatrix} a_1 \\ a_2 \\ a_3 \\ a_4 \end{bmatrix} = \begin{bmatrix} \text{Estimated feature defects present} \\ \text{Regression test case execution rate} \\ \text{Estimated test case defects present} \\ \text{New test case execution rate} \end{bmatrix}$$

with its local output vector given as

$$b^{dd} = \begin{bmatrix} b_1 \\ b_2 \end{bmatrix} = \begin{bmatrix} \text{Estimated failures due to feature defects} \\ \text{Estimated failures due to test case defects} \end{bmatrix}$$

$$= \begin{bmatrix} \mu_f a_1 (a_2 + a_4) \\ \mu_t a_3 a_4 \end{bmatrix}$$

where μ_f gives the probability that a test case will discover a defect and μ_t gives the probability that a test case will reveal its own internal defect. Note the appearance here of two copies of the defect encounter model of Eq. (10). For simplicity, we assume that regression tests are generally correct, having been successfully executed on a previous release of version of the product; hence, only new test execution provides the "predator" for finding test case defects. This simplification is not necessary, but the alternative requires another means for estimating the defects present in the currently active subset of the regression test suite (i.e., perhaps another queue on the failure analysis activity, or an external input.) Note that the defect estimation model of Section 2.8 embeds naturally into the above to provide a single algebraic defect detection component with six inputs and two outputs.

2.10. *An Algebraic Model of the Proportional Splitting Component*

The conditions given in Eq. (11) yield the following equations for the proportional splitting component. The piecewise form is chosen to maximize numerical stability and to avoid requiring a regularizing term in the denominator for cases where the references are near (or at) zero. For an input vector defined as

$$a^{ps} = \begin{bmatrix} a_1 \\ a_2 \\ a_3 \end{bmatrix} = \begin{bmatrix} \text{Reference R1} \\ \text{Reference R2} \\ \text{Source} \end{bmatrix}$$

the algebraic form of the proportional splitting component is given as

$$b^{ps} = \begin{bmatrix} b_1 \\ b_2 \end{bmatrix} = \begin{bmatrix} \text{O1 (Fraction of Source)} \\ \text{O2 (Fraction of Source)} \end{bmatrix}$$

$$= \begin{cases} \begin{bmatrix} ((a_3 a_1)/(a_1 + a_2)) \\ a_3 - ((a_3 a_1)/(a_1 + a_2)) \end{bmatrix} & \text{if } a_1 > a_2 \\[2ex] \begin{bmatrix} a_3 - ((a_3 a_2)/(a_1 + a_2)) \\ ((a_3 a_2)/(a_1 + a_2)) \end{bmatrix} & \text{if } a_2 > a_1 \\[2ex] \begin{bmatrix} (a_3/2) \\ a_3 - (a_3/2) \end{bmatrix} & \text{if } a_1 = a_2 \end{cases}$$

3. Assembling the Model

In the preceding sections, we have defined each component of the modeling framework; we now consider the process of composing them to represent a specific software development process. We begin with the specification of the workflows illustrated in Fig. 1. Recall from Fig. 3 that the partition L_{11} of the interconnection matrix specifies the linear feedback from subsystem outputs to the subsystem inputs. The partition L_{12} defines how the external inputs enter into the subsystem inputs. Given that the inputs and outputs of each subsystem are vectors, let us further partition L_{11} into subpartitions L_{11ij}, and L_{12} into subpartitions L_{12i} satisfying $a_i = \left(\sum_{j=1}^{N} L_{11ij} b_j \right) + L_{12i} u;\ \forall i \in \{1, 2, \ldots, N\}$. That is, given N subsystems, let L_{11ij} specify the linear feedback from the output of the jth subsystem to the input to the ith subsystem for all pairs of subsystems, $\langle i, j \rangle$. Hence, the subpartition L_{11ij} is an $m \times n$ matrix, given that a_i is $m \times 1$, and b_i is $n \times 1$. Likewise, the L_{12} partition is divided into subpartitions of size $m \times p$, where u is $p \times 1$, representing the contribution of the external input, u, to the local subsystem input of any component. With this interpretation of the subpartitions of the matrix L, it is clear that the placement of a nonzero element into a subpartition L_{11ij} specifies a scaled flow from an element of the local output vector of subsystem j, to an element of the local input vector of subsystem i. For clarity, we recall that the scale factors selected for insertion into the interconnection matrix specify the translation from the work items completed, as evident in the output vector, to the number of by-product work items created as a result.

To compose a model of a particular development process and/or project, one must first identify the subsystems that compose the process, and then specify workflows that connect them. Appendix A details the development of a full interconnection matrix corresponding to Fig. 1.

4. A Simulation Study

The purpose of this study is to examine the behavior of the model under a typical software development scenario. The simulation method is briefly described, and then the example software development scenario is detailed; a summary of the simulation results follows.

4.1. *Simulation Method*

Interconnected system-of-systems models, as illustrated in Fig. 3, can be simulated by the following modification of the improved Euler method.

Let h be the size of a simulation time-step. Given an initial state x_t and external inputs, u_t, taken as constant over the interval $[t, t + h]$, compute $m_t = f(x_t, a_t)$ as an estimate of the time derivative of the system state at time t, i.e., $\dot{x}_t = f(x_t, a_t)$ (if a_t is not available, one can solve for it using

x_t and u_t). Then $\hat{x}_{t+h} = x_t + hm_t$ is an estimate of the future state at time $t + h$ as generated by the first-order truncation of the Taylor series expansion of x.

One may then solve for the subsystem input and output vectors consistent with the estimated future state, and expected external input at time $t + h$. That is, one may solve for a_{t+h} satisfying $a_{t+h} = L_{11}g(\hat{x}_{t+h}, a_{t+h}) + L_{12}u_{t+h}$. In the general case, such a solution may not exist, or may not be unique. The Implicit Function Theorem provides a test to establish the existence of a solution. As will be shown below, an implicit solution is not necessary for the special case of models built within the framework developed above; an explicit calculation of the unique solution is possible.

Upon obtaining a_{t+h}, $m_{t+h} = f(\hat{x}_{t+h}, a_{t+h})$ can be computed as an estimate of the time derivative of the system state expected at time $t + h$. Here, the function f is again applied to generate the time derivative, in the manner specified by the state evolution equations, given the estimate of the future state, $\hat{x}_{t+h}$, and the estimated future subsystem inputs, a_{t+h}. An estimate of the time derivative at each end of the simulation step is now available, as required by the improved Euler method; hence, one computes the improved Euler step by averaging the two derivative estimates, and applying the average over a time-step to move from the current state to the next state, i.e., $x_{t+h} = x_t + \frac{h}{2}(m_t + m_{t+h})$, classically referred to as the trapezoidal rule.

Listing 1. Improved-Euler Step Adapted for System-of-Systems Composed Models

```
01.   Compute  m_t = f(x_t, a_t)
02.   Compute  x̂_{t+h} = x_t + hm_t
03.   Solve  a_{t+h} = L_11 g(x̂_{t+h}, a_{t+h}) + L_12 u_{t+h}
04.   Compute  m_{t+h} = f(x̂_{t+h}, a_{t+h})
05.   Compute  x_{t+h} = x_t + (h/2)(m_t + m_{t+h})
06.   Update  t = t + h
```

As mentioned for Step 03, Listing 1, the implicit solution to find the a_{t+h} and b_{t+h} vectors consistent with a given $\hat{x}_{t+h}$ and u_{t+h} is not necessary when working with models built within the current framework. Figure 11 shows the categories of data flow present among the various data vectors computed during the modified Improved Euler step described above. All upward connections pass through integration operations — hence

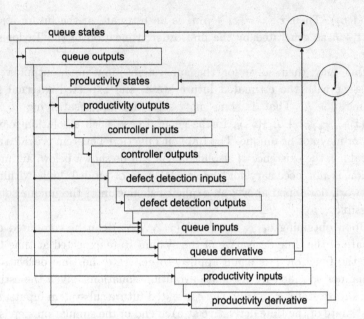

Fig. 11.   Allowed data flow relations among the classes of sub-component data vectors.

no amount of change in a vector in the lower category can induce a change in any vector of the upper, at the current time instant. Given the absence of direct cycles in the data flow graph, there is an explicit solution to the problem of finding the subsystem inputs and subsystem outputs given the system state and global system inputs. Specifically, most of the component outputs, b_{t+h}, are identified with system states — from this set of outputs, direct computation of the subsystem inputs needed to determine the remaining subsystem outputs, and thus the entire vector, b_{t+h}, is possible. Therefore one can directly compute $\hat{a}_{t+h} = L_{11}\hat{b}_{t+h} + L_{12}u_{t+h}$ without having to solve implicitly defined non-linear/linear equations.

4.2. *Simulation Results*

In order to demonstrate the behavior of the model, nominal parameter values have been chosen to generate an arbitrary instance of the model. The resulting simulation traces are described below. The choice of parameters is guided by the following hypothetical software development project executed under an incremental development process: the project entails two serial increments; the first increment is composed of 10 work items for feature coding, 70 work items for test case creation, and 70 work items for regression test execution; the second increment will add 20 work items for feature coding, 130 work items for test case creation, and 130 more work items for regression testing. There are 29 (workforce) FTEs allocated to the project; five are assigned to each development activity, except for the feature and test case correction activities, to each of which only two are allocated. As above, it is assumed that the test cases for a given increment cannot be executed until the feature coding work on the increment is complete.

Simulation of this scenario yields 79 input and output traces from the various modeling components. The remainder of this section details the interesting features of the simulation traces and relates them to the earlier model development. By convention, the legend lists inputs before outputs; also two sets of y-axes are used throughout: the left for work item *counts*, and the right for work item *rates*. Whether a given trace is plotted with respect to the left or the right axis is readily discernable from the legend; the reader should note that accumulations of rates over time (e.g., the cumulative sub-model outputs) become counts. The parameters of the simulation, and a discussion of their selection, can be found in Appendix B.

4.2.1. *Feature Coding*

Figure 12 shows the inputs and outputs of the productivity model for the feature coding activity. Initially, the workforce ramps up to process work items, reaching a peak of productive capability at week 4. As the remaining work items (i.e., workload size) reduce to zero, and the interesting challenges wane, the productive *capability* (as distinct from *productivity*) declines to a steady state value around week 13.

Recall that cumulative productive capability is the integral of productive capability over time. Consequently, by the non-zero steady-state productive capability after week 13, the cumulative capability increases at a constant rate, despite there being no actual work items to process. By taking the difference between this capability, and the cumulative out-flow of the feature coding queue (see Fig. 13), we obtain a metric for cumulative wasted effort, as detailed in Fig. 14.

Figure 13 shows the I/O traces for the queue simulation of the feature coding activity. Note near week 6 that the cumulative out-flow and the queue level traces cross at a value of 15 work items. This is a natural consequence of the zero in-flow rate, which leads to queue-level and cumulative out-flow traces that are mirror images; the crossing at a level of

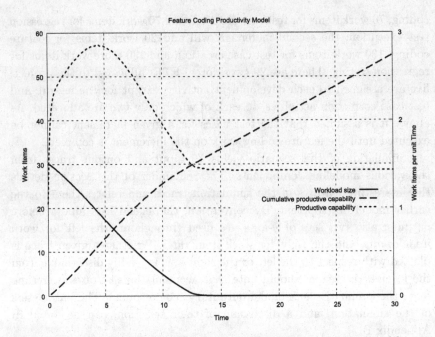

Fig. 12.   Simulation trace of I/O for the feature coding productivity model.

15 work items is therefore logical, being the half-way point of completing the 30 initial work items. The out-flow rate closely tracks the commanded out-flow rate until near week 12. At this point the queue level has dropped sufficiently to cause the saturation function embedded within the queue equations to begin its transition from 1 to 0. Recall that the saturation function acts as a switch; in this case moving the queue from a mode that honors the commanded out-flow rate, into a mode that tracks the minimum of the in-flow rate (which is zero in this instance) and the commanded rate. The *queue level* in Fig. 13 and the *workload size* in Fig. 12 are quite similar, as are the *productive capability* of the productivity model, and the *commanded out-flow rate* of the queue. They are similar because, for each pair, one is the source of the other. This is precisely the data dependency illustrated in Fig. 2.

Given the simulation's hypothetical development scenario, it may seem questionable that the queue level for the feature coding activity begins with 30 work items, rather than just the 10 to be completed for the first release. This method of tracking the future work for the feature coding team allows the impending future tasks to influence the workload-dependent

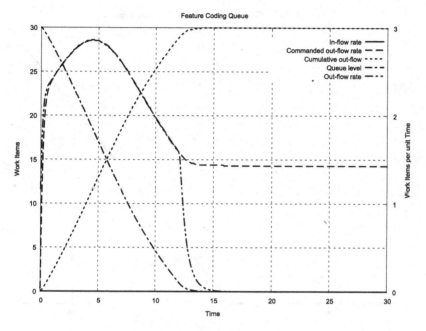

Fig. 13.   Simulation trace of I/O for the feature coding queue.

productivity equations. For a short project, this may be a relevant choice. If such is undesirable, an algebraic component can be added to control the in-flow to the feature coding queue as a function of the feature coding cumulative out-flow. Such a control could, for example, inject new work items as the completion of the features for the first release nears.

4.2.2. *Test Case Coding*

The discussion of the test case coding activity largely follows that of the feature coding activity. Figures 15 and 16 reveal the same data dependency between the inputs and output of the productivity and queue sub-models as discussed in Section 4.2.1. The productivity model quickly ramps up to a peak at 5 weeks, and decreases until a steady state is reached near week 26. Again, all 200 work items are initially present, rather than just the 70 that correspond to the first increment. The crossing of the queue level and the cumulative out-flow at week 11 is slightly offset from the 100 work item midpoint due to the small, but non-zero, in-flow rate (driven by the failure analysis activity, as illustrated by workflow 9 in Fig. 1). The non-zero in-flow rate is also the reason that the cumulative out-flow at

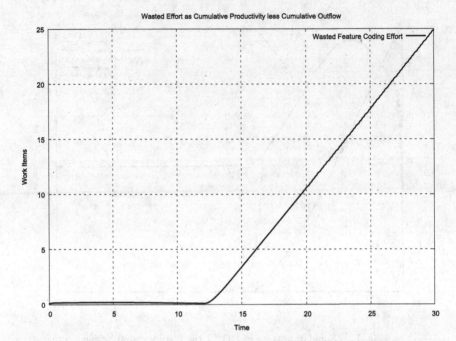

Fig. 14.   The "wasted effort" metric corresponding to the feature coding activity traces in Figs. 12 and 13, computed as the difference between the cumulative productive capability of the feature coding productivity model, and the cumulative outflow of the feature coding queue.

week 30 (Fig. 16) is larger than the initial 200 work-items (due to rework). As with the previous activity, the queue level vanishes sufficiently near time 26 to cause the saturation function within the queue to switch modes (i.e., to track the minimum of the commanded out-flow rate and the in-flow rate, rather than simply to track the commanded out-flow rate), as is seen in Fig. 16 by the rapid reduction of the out-flow rate beginning at that time.

As mentioned above, the failure analysis activity contributes a non-negligible in-flow rate to the test case coding activity. In constructing the simulation, we have hypothesized (as per Fig. 1) a learning process during the failure analysis activity. That is, during the course of uncovering the cause of initial test failures, various shortcomings may be identified in the general testing strategy, leading to augmentations of the test strategy, and the definition of new test case specifications for the test case coding activity. The two "humps" in the in-flow rate trace correspond to the processing

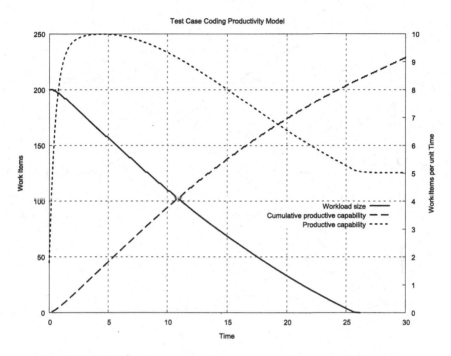

Fig. 15. I/O Simulation trace for the test case coding productivity model.

of the two increments in the hypothetical development scenario — in particular, the rising edge of each hump can be shown to correspond to the commencement of testing on each increment.

4.2.3. *New Test Case Execution*

The new test execution activity differs significantly from the previous simulation traces. In Fig. 18, note the contrast between the slow build-up of the productive capability exhibited by the new test execution productivity model when compared to the ramp-up of feature coding productivity (Fig. 12) or test case coding productivity (Fig. 15). To understand the evolution of the productive capability, recall the state evolution equation for the productivity model, given in Eq. (1). Figure 17 plots the field defined by Eq. (1) in conjunction with the model parameters used in the simulation study. This field describes the time derivative of the productive capability for the new test execution activity as a function of the current productive capability and the current workload size. Any portion of the

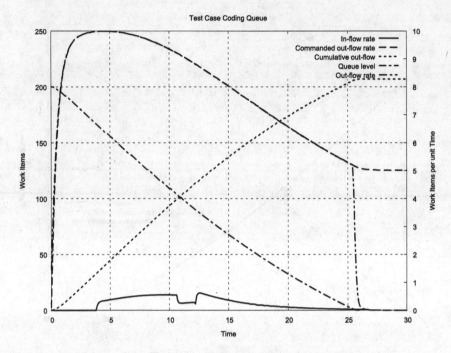

Fig. 16.   I/O Simulation trace for the test case coding queue.

field above the zero plane represents an acceleration of the productive capability; correspondingly, any point below the zero plane represents a deceleration.

Because the value of the field at a given' coordinate is the time derivative of the productive capability, and one of its parameters is productive capability itself, the intersection between the field and the zero plane indicates the natural equilibrium point of the productivity model corresponding to a given workload size, holding all other parameters constant. While sub-figure (a) indicates that an increase in the current workload size will increase the productive capability at which the equilibrium point is reached, sub-figure (b) indicates that there is a limit to the potential productivity increase achievable by this method.

The productive capability derivative values encountered during the simulation are plotted on the field; each point on the trace corresponds to one week of the simulation. The trace in Fig. 17, sub-figure (a) begins with zero workload, and zero productive capability. Note that an increase in the initial workload would significantly increase the initial rate of growth

(a)

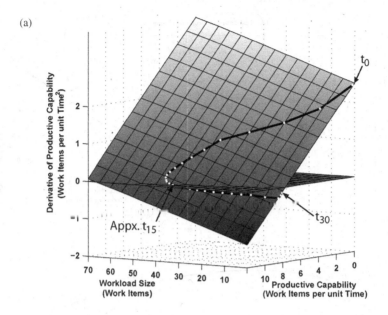

(b)

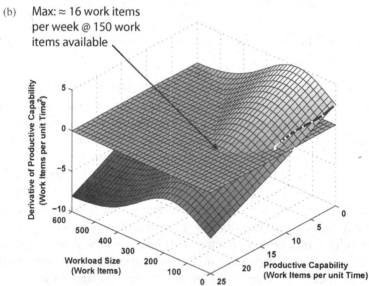

Fig. 17.   (a) Trace of the productive capability derivative for the new test case execution activity, plotted on the field defined by the state evolution equation for the same (Eq. (1)), using the example simulation parameters. The zero plane is plotted for reference, and each point on the trace represents one simulation week. (b) The same plot from a wider view of the field. Note the clear maximum sustainable productive capability given the current parameters.

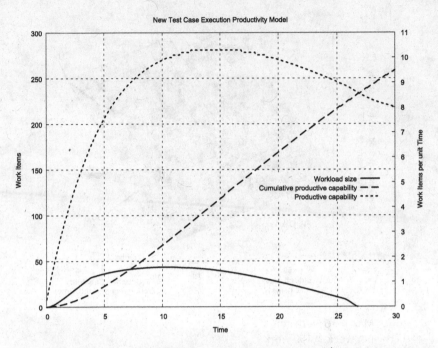

Fig. 18.   Simulation trace of I/O for the new test case execution productivity model.

of productive capability — this is part of the explanation for the slow initial ramp up of productive capability for the new test execution activity when compared to the coding activities: the coding activities begin with a non-zero workload. The other factor is the lack of a sufficient in-flow rate to drive the workload size into the range required for peak productivity. From approximately week 13 through week 17, the derivative trace lingers near the zero plane; this corresponds to the peak region of the productivity curve in Fig. 18. The preceding discussion illustrates that the work item in-flow rate to a development activity has a significant impact on the activity's ability to maintain peak productivity. In fact, the in-flow rate must match the peak productive capability if the peak productive capability is to be sustained over a duration, ignoring the potential complicating factor of coordination constraints.

Figure 19 shows a large in-flow rate to the new test execution activity, peaking at 10.5 work items per unit time near week 6. Figure 20 decomposes the peculiar shape of the in-flow rate trace into its constituent contributions; the summation of these components occurs in the linear feedback within

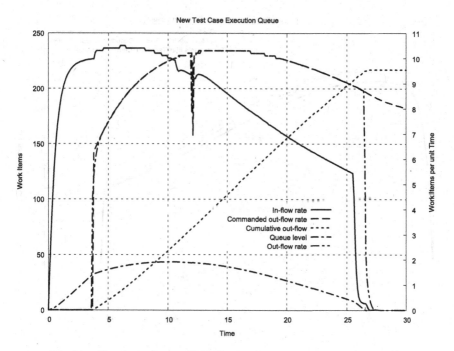

Fig. 19.   Simulation trace of I/O for the new test case execution queue.

the system of systems framework. The in-flow vanishes near week 27, as the
outflows from its "upstream" activities cease.

The entire body of work items to be completed by the new test
execution activity arrives through the in-flows; from weeks 0–4, work items
simply accumulate because the commanded out-flow rate of the queue is
held at zero, preventing any work item processing. After week 4, work-
item processing begins, causing a visible knee in the queue level plot, and
beginning the ascent of the cumulative out-flow trace. The queue level for
the remainder of the simulation reflects the balance between the active in-
flow and out-flow of the queue: The queue level reaches a maximum just
after week 10; i.e., the derivative of the curve is zero precisely when the
in-flow rate and the out-flow rate traces are equal. Unlike the preceding
analyses, and due to the initially empty status of the work item queue,
there does not appear to be a meaningful interpretation of the intersection
of the cumulative outflow, and the queue level traces.

Near time 12, there is an unusual downward spike in the trace of
the commanded out-flow rate, and consequently, the actual out-flow rate.

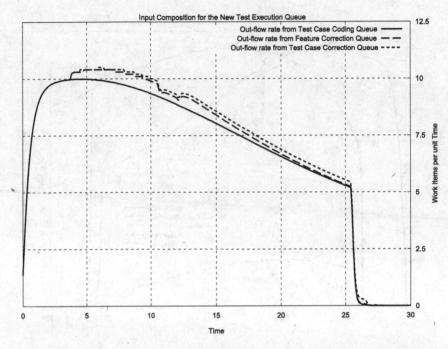

Fig. 20. Area graph of the test case execution activity's in-flow rate composition.

This is the first time in the analysis of the example simulation that the commanded out-flow rate supplied to the queue deviates from the productive capability of the activity's productivity model; this is due to the action of the scheduling controller regulating the cumulative progress of the new test case execution activity (by lowering the commanded rate) as the new test case execution progress (i.e., cumulative out-flow) approaches the allowed threshold. Two factors prevent the scheduling controller from stopping new test case execution entirely (i.e., driving the out-flow rate to zero) in this instance: (i) feedback from the correction activities (described below) provides an in-flow of work items that are not subject to coordination constraint being enforced by the controller, and (ii) the feature coding team completes the second increment just after week 12, thereby satisfying the coordination constraint and allowing new test execution activity to continue (i.e., raising the threshold on the maximum allowed progress). Figure 21 shows the threshold that is being enforced on the progress of the new test case execution activity by the scheduling controller. Near time 12, the controlled cumulative out-flow

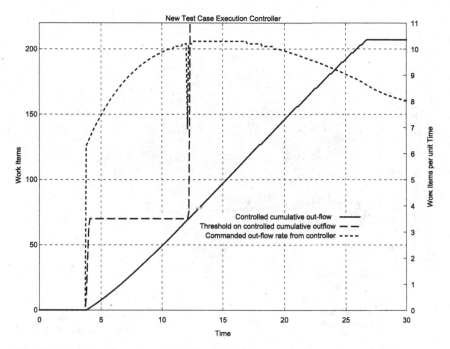

Fig. 21. Relation between threshold, controlled cumulative outflow, and the commanded rate regulated by the controller.

nears the threshold, hence the scheduling controller's action to reduce the commanded out-flow rate; shortly thereafter, the threshold rises, and execution continues.

Returning to Fig. 19, shortly after week 26, one observes the now-familiar transition from tracking the commanded out-flow rate, to tracking the minimum of the in-flow rate (zero, at that time) and the commanded rate as observed above. At the end of the simulation, the cumulative out-flow of the activity queue levels off near 213 work items — a value higher than even the cumulative out-flow of the test case coding activity. This is due to the feedback loops providing "re-work" from the correction activities (described below).

4.2.4. *Regression Test Case Execution*

In Fig. 23, the productive capability of the regression test execution activity peaks near week 10. While this peak is reached more quickly than the productivity ramp-up for the new test execution activity (Fig. 18), it is

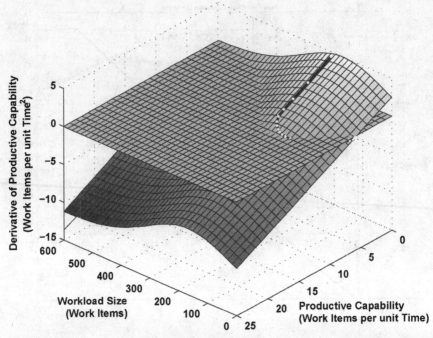

Fig. 22. Trace of the productive capability derivative for the regression test case execution activity, plotted on the field defined by the state evolution equation for the same (Eq. (1)), using the example simulation parameters. The zero plane is plotted for reference, and each point on the trace represents one simulation week.

slower than for either the feature or test case coding activities (Fig. 12 or Fig. 15, respectively.) Figure 22 shows the derivative of the productive capability for the regression test case execution activity, in the context of the field defined by its state evolution equation in Eq. (1), using the same parameters as the example simulation. The first four weeks of the simulation drive the productive capability upward with no impact on the workload size. This is due to a coordination constraint requiring the progress of the regression test execution activity to remain at zero until the first release is complete; that is, the workforce prepares for the upcoming work, but cannot actually start. The first release is ready at week 4, and hence the normal evolution of the productive capability begins, depleting the workload size, and sweeping the derivative trace rightward, again as illustrated in Fig. 22.

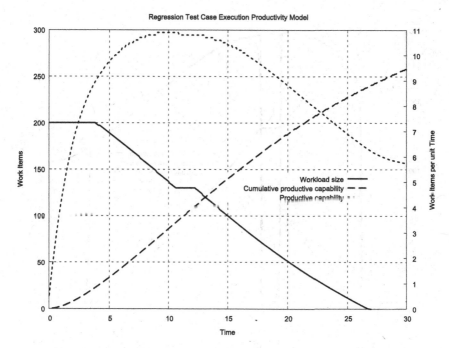

Fig. 23.   Simulation trace of I/O for the regression test productivity model.

Returning to Fig. 23, the shape of the "workload size" trajectory for the regression test execution activity exhibits two horizontal segments, one over weeks 0 through 4 and the other over weeks 10.5 through 12. As mentioned in the preceding discussion, these horizontal segments are due to the action of the scheduling controller for the regression testing activity. Particularly, the horizontal region starting a 10.5 weeks demonstrates the full operation of the scheduling controller to halt the ongoing progress of a development activity in order to preserve a coordination constraint. Note that this later halting of progress is not detectable in Fig. 22 because both the productive capability and workload size remain nearly constant; hence, the points on the trace during the period simply overlap.

As with the new test execution activity (i.e., Fig. 19), Fig. 24 shows that there is no progress made by the regression test execution activity until time unit 4, despite the fact the activity has a significant workload size. This delayed start is due to the scheduling controller, as can be verified by noting the rising edge of the commanded out-flow rate at time 4 — recall

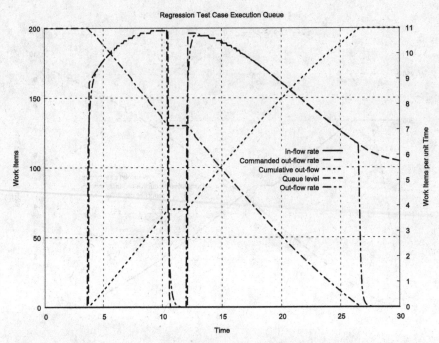

Fig. 24. Simulation trace of I/O for the regression test activity queue. The inflow rate remains at zero.

that regulating the commanded rate is the schedule controller's actuation mechanism for controlling a development activity's ability to make progress.

Figure 24 also shows that the commanded out-flow rate is reduced to zero between time units 10.5 and 12, with the actual out-flow rate quickly tracking zero in response. The explanation is that near time 10.5 the regression test execution activity exhausts its initial first increment workload of 70 test cases. Hence the scheduling controller forces the regression testing activity to wait for the availability of the next increment, which occurs near week 12. Consequently, at week 12, the commanded output (closely tracked by the output rate) rises again to the full productive capability given by the productivity model. Looking back to Fig. 13, one may verify that this is indeed the time of completion for the second increment.

Because the in-flow rate to the regression test execution activity is zero, the queue level and cumulative out-flow traces are mirror images, crossing at week 15, with a level of 100 work items — the midpoint of the initial 200

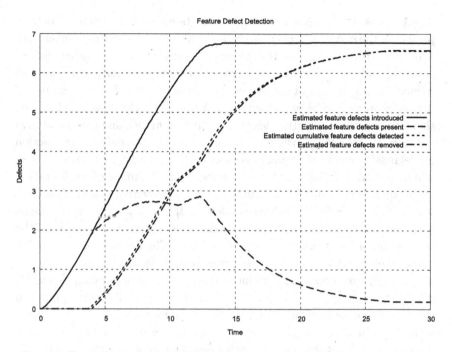

Fig. 25. Simulation traces for the feature defect introduction and detection models.

work items. Lastly, at approximately time 27, the regression test activity removes the last work items from its queue, and the out-flow rate falls to zero (as the minimum of the commanded rate and the in-flow rate).

4.2.5. *Defect Introduction, Defect Detection, and Failure Analysis*

Defect introduction and detection are the implicit activities that drive the failure analysis activity. For the example simulation, Fig. 25 shows the tracking of estimated feature defects within the model. The estimated number of feature defects introduced into the code is given by the COQUALMO-based defect introduction model described in Section 2.8. The introduction of feature defects begins at week zero, corresponding with the commencement of feature coding, and continues as long as feature coding progresses, reaching a steady state value of 6.8 feature defects introduced as feature coding comes to a completion near week 14. Note that the parameters chosen for the defect introduction model are quite optimistic; any non-example scenario will likely expect orders of magnitude higher defect introduction behavior.

At week 4 the trace recording the instantaneous estimate of defects present in the software departs from the total number of defects introduced. This departure is due to the commencement of testing, hence failure analysis, and ultimately, defect correction (discussed later), which acts to reduce the population of defects in the software. Literally, the number of defects estimated to be present in the feature code is the difference between the output of the cumulative defect introduction model, and the cumulative out-flow of the defect correction activity. Hence, the defect introduction model must predict the number of defect correction work items, rather than a literal count of defects, and can therefore take advantage of the literature on the occurrence rates for defects in the various severity classes, if desired. The upward turn in the estimated defects trace at week 11 corresponds with the pause of the regression testing activity that occurs with the completion of the increment 1 regression tests. The feature coding activity pushes defect introduction higher while producing increment 2 features, but the test execution rate is reduced (i.e., only new tests being executed) so defects accumulate. The downward turn at week 12.5 corresponds to the completion of feature coding for the second increment of the example scenario, which allows regression testing to resume, leading to defect discoveries and, in turn, feature corrections. The remaining traces in Fig. 25 plot the cumulative defects detected and the cumulative defects removed. Examining the slope of the trace of cumulative defects detected, one may confirm that the upward turn in the estimate of remaining defects is a consequence of the slowdown in testing, and consequent slowdown in defect detection.

The parameters selected for the failure analysis activity and the feature correction activity in this example simulation yield an extremely short duration between defect detection and removal. Through calibration, the development activity models can represent development processes with a significant failure analysis and/or correction phases — this has simply not been done in this example. Note, lastly, that the detection of defects is asymptotic to zero, as seen in the trace of estimated defects in the feature code after week 12.5, where the detection, analysis and correction steps are the only active activities. This asymptotic behavior matches the conventional wisdom that testing will never reveal all defects.

Figure 26 shows the simulation traces related to tracking the estimated defects in the set of test cases that have been produced. The total number of defects introduced increases steadily until just after week 26, which corresponds to the completion of the test case coding for the second

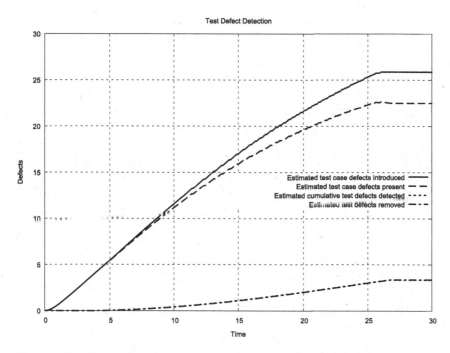

Fig. 26. Simulation traces for the test case defect introduction and detection models.

increment, as shown in Fig. 16. Figure 19 and Fig. 24 show that the completion of test case execution also occurs in weeks 26–27, hence the cumulative number of defects detected levels off. One unusual feature of Fig. 26 is the extreme proximity of the cumulative test case defect detection trace, and the cumulative removal trace; in the plot, they are indistinguishable. Interpreted in the context of a software process, such would imply that the correction of test case defects is instantaneous; in this case it simply implies that the parameters chosen for this example simulation represent an unrealistic case. As with Fig. 25, the estimated defects present in the new test cases is given as the estimated defects introduced, less the number removed and hence the trace levels off shortly after week 26 as well.

Figure 27 gives the simulation traces for the productivity model of the failure analysis activity. Again, the example parameters configure the failure analysis activity model to process work items very quickly, yielding a sustained near-zero workload size, and a correspondingly near-constant productive capability. The simulation traces for the failure analysis queue

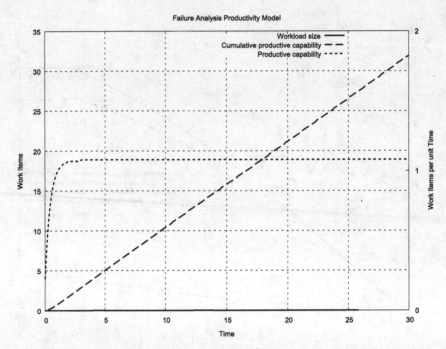

Fig. 27.   Simulation traces for I/O of the productivity model of the failure analysis activity.

model in Fig. 28 illustrates the unusual case where the out-flow rate predominantly tracks the in-flow rate (as the minima of the in-flow rate and the commanded rate, given the empty queue), i.e., passing all in-flow straight through to the out-flow, as expected to best satisfy the demand under insufficient supply. Near week 11, there is a sharp downward turn in the in-flow rate to the failure analysis queue; this corresponds to the regression testing activity reaching its threshold of work items to be run on the first increment of the example scenario. Hence, there are fewer tests being executed to drive defect detection, and the in-flow rate to failure analysis falls to just the defect detection rate from the new test case execution activity. Just after week 12, there is another downward spike, which immediately turns upward; this corresponds to the new test case execution activity reaching it's first increment threshold just as the feature coding activity completes the second increment, as described Section 4.2.3. Thus the in-flow rate rises to a new peak, near 13 weeks, driven by the defects detected by both regression and new test case execution on the second increment features and tests. The slight deviation between the

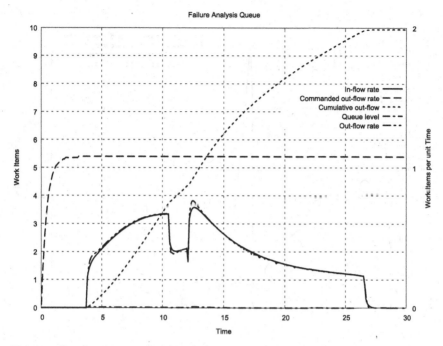

Fig. 28. Simulation traces for I/O of the queue model for the failure analysis activity. This plot shows the sum of the inputs to the two queues (§2.7) maintained for the failure analysis activity; i.e., the view of the project manager, who cannot know *a priori* which failures originate from feature defects, and which from test defects.

in-flow rate and the out-flow rate is due to the gain of the control logic embedded within the queue model (as configured through the time constant, τ, in Eq. (7)) in conjunction with the relatively large time-step used in the simulation. Near week 27, the last test cases are executed, and hence defect detection (and the in- flow rate it generates for failure analysis) vanishes, leaving nearly 22 test case defects undiscovered (see Fig. 26).

Figure 29 shows the inputs to the proportional splitting component (see Section 2.7). In the configuration used for this simulation example, these inputs are driven directly from the queue levels of the two queues of the failure analysis activity (see Fig. 10). The high rate of progress made by the failure analysis activity yields very small values for the levels of the failure analysis queues; this in turn leads to a high volatility in the output of the proportional splitting component in Fig. 30. The splitting component is sensitive to the ratio of the inputs, which changes rapidly for small input magnitudes. Note that Fig. 30, plotting the outputs of the

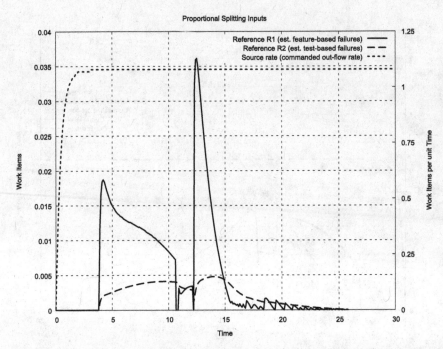

Fig. 29.   Inputs to the proportional splitting component of the failure analysis activity.

proportional splitting component, shows how the source rate is divided into two channels, congruent to the magnitude of the inputs, R1 and R2. Hence, the sum of the proportional splitting outputs is equal to the source rate. The symmetry in Fig. 30 is a consequence of the relatively constant source rate: As one proportion grows, the other must shrink proportionally. Between weeks 0 through 4, the proportions are evenly split at 50%, however the symmetry is broken by the rising source rate.

In Fig. 31, it is clear that the volatility of the commanded rates generated by the proportional splitting component is damped by the failure analysis queues (indeed, the queues *track* the commanded rate rather than honor it precisely). Note in Fig. 31, that the cumulative failure analysis out-flow matches that shown in Fig. 28, and that it is simply the sum of the cumulative out-flows from the activity's two queues.

4.2.6. *Feature Correction*

In Fig. 32, one sees that the feature correction productivity model is configured to generate a productive capability of just over 0.5 work items

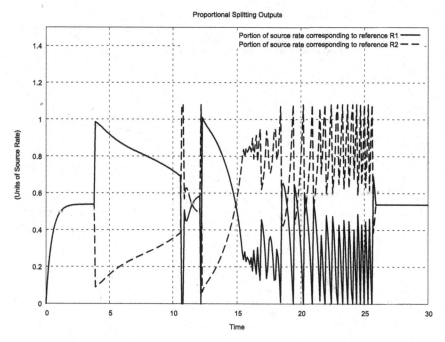

Fig. 30. Output traces from simulation of the proportional splitting component of the failure analysis model.

per week when its workload size is approximately zero work items (which is the present case). Because the example scenario for this simulation places no coordination constraints on the feature correction activity, this productive capability passes, unmodified, through the scheduling controller and enters as the commanded rate to the feature correction queue in Fig. 33. Recall that the in-flow rate to the feature correction queue is driven (through workflow 10 in Fig. 1) by the rate of *feature change request* work items produced by the failure analysis activity in response to determining that the cause of a failure is a defect in the feature code. The in-flow rate trace in Fig. 33 exhibits some low amplitude jitter in weeks 16 through 22 due to the volatility in the proportional splitting component that apportions productive capability to the failure analysis queues (Fig. 30). As described in Section 4.2.5, this jitter is too small to be visible in the cumulative out-flow (i.e., the integral of the out-flow rate) plotted in Fig. 31, but is sufficiently large to be visible in the in-flow to the feature correction activity.

In Fig. 33, the out-flow trace generally tracks the in-flow rate, except during weeks 13–14, where it trends toward the commanded rate. This trace

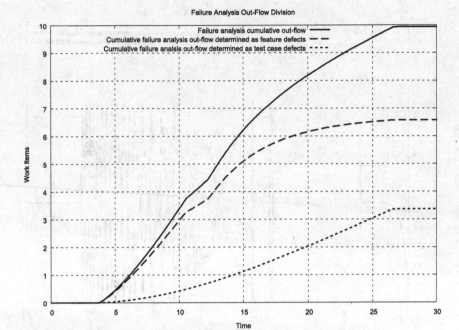

Fig. 31. Simulation trace of the cumulative out-flows from the two failure analysis activity queues, whose commanded rates are driven by splitting the failure analysis productive capability into the same relative proportions as the queue levels.

illustrates the tolerance introduced through the smoothed min function, embedded into the tracking logic of the queue state evolution equation (Eq. (7)). Per Eq. (8), the smoothed min prefers its first parameter, returning the second parameter only if it is larger than the first by a difference of at least 1. Over the unit-width interval where the second parameter is larger, but with a difference less than 1, the smoothed min returns a linear combination of its parameters. Returning to the out-flow rate trace in Fig. 33, because the queue level is nearly zero, we expect the out-flow rate to track the minimum of the in-flow rate and the commanded rate; but from weeks 7–11 and 13–14, the out-flow rate tracks the in-flow rate. Note that these traces are rates, and are therefore measured on the right-hand axis; the difference between the in-flow rate and the commanded rate is less than 1/10 of the difference that would be required to cause the smoothed min to transition completely to the commanded rate.

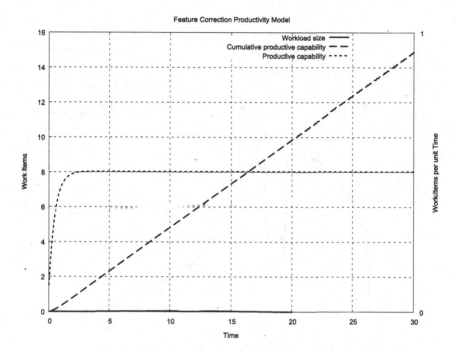

Fig. 32.   Simulation traces of I/O for the feature correction productivity model.

Near week 10.5 the out-flow rate deviates from the in-flow rate due to the combination of the simulation time step size, the gain of the control logic that tracks the smoothed minimum of the in-flow and commanded rates, and the steep descent of the in-flow rate that is being tracked (the gain is configured through time-constant τ in Eq. (7)). A curious aspect of Fig. 33 is that the out-flow rate does not track the in-flow rate as closely over weeks 12–13 as it does over weeks 7–11 (both are cases where it should instead be tracking the commanded rate). It seems that three factors are responsible: (i) the in-flow rate quickly falls off again, leaving little time for the out-flow rate to track it; (ii) the peak at week 11 is nearly twice as far above the commanded rate as that at week 9, hence, the rate being tracked will contain a larger proportion of the commanded rate in the linear combination produced by the smoothed min function; and (iii) the modified Improved Euler simulation method (see Section 4.1) averages between the current derivative, and the derivative expected at the state resulting from a Newton step based on the current derivative. Therefore the sharp decline

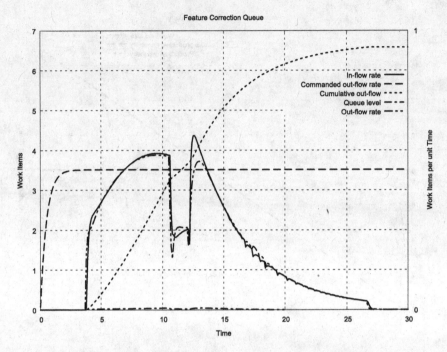

Fig. 33. Simulation traces of I/O for the feature correction queue model.

of the in-flow rate may have dampened the tracking indirectly through the simulation method.

4.2.7. *Test Case Correction*

In this example simulation, the productivity model for the test case correction activity is configured identically to the feature correction activity's productivity model. Hence, Fig. 34 is nearly identical to Fig. 32 (save for small variation in the workload size).

Figure 35 shows another case where the workload size remains at zero largely due to the small in-flow rate. The out-flow rate therefore tracks the in-flow rate rather than the commanded rate, as described in Section 2.3. The small dip in the in-flow rate seen at week 12 directly corresponds with the evanescent pause in the new test execution activity (Fig. 19) because only the execution of new tests can reveal the defects introduced in the new tests, thereby providing work items to failure analysis, which eventually passes the change request to test case correction.

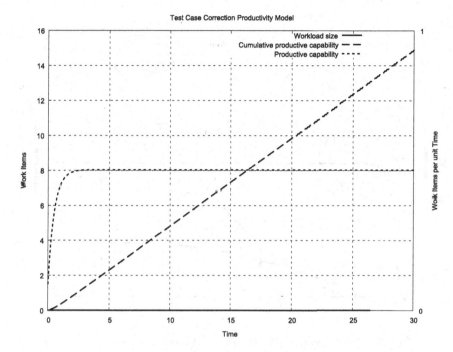

Fig. 34.   Simulation trace of I/O for the test case correction productivity model.

The small bump in the out-flow rate at week 27 seems to be another artifact of the simulation method generated by a tiny rise in the in-flow rate at that time.

5. Model Calibration

In this section, a distinction is made between ratio and interval scale data to call out that calibration can still be performed even under an unknown uniform translation of the calibration data (e.g., a constant measurement error). In an informal sense, if one assumes one has ratio scale data, it implies that one has accurate (but perhaps noisy) measurements.

In defining the calibration routines, we have generally assumed that the matrix L in Fig. 3 has a partition L_{21} that is an identity matrix whose dimensions match the length of the subsystem output vector, and that the partition L_{22} consists entirely of zeros. That is, the subsystem outputs *are* the external outputs of the system.

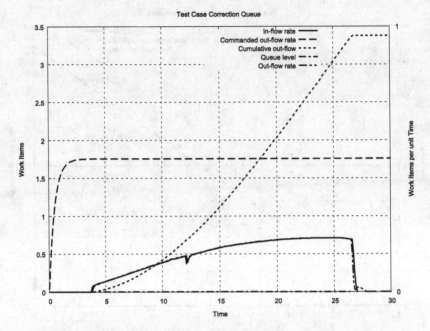

Fig. 35.   Simulation trace of I/O for the test case correction queue model.

While the calibration routine for ratio scale data is consistent with traditional practice, the routine for interval scale data explores alternatives for constructing novel calibration routines.

5.1. *Calibrating with Ratio Scale Data*

Empirical observations are assumed to be spaced sparsely over a regular time grid, $G = \{t_0 + kh : k = 0, 1, \ldots\}$, and hence will not capture the external inputs for every point on the grid. Because it is necessary to know the value of the external inputs at each point on G, we assume that the external inputs are piecewise constant, and that the grid points at which they are updated are recorded, hence the external inputs present at any grid point can be reconstructed. As a natural consequence, a zero-order hold constraint on the external inputs is satisfied. Recall that, for the modeling components described in Section 2, the general subsystem outputs, b, can be explicitly computed given only the system state vector, x, and the external input vector, u (see Section 4.1). One can therefore state that, for models built within the preceding framework, there exists an explicit function $b = \tilde{g}(x, u)$. For general composed models, this result requires certain conditions to be satisfied according to the implicit function theorem.

Given a set of N empirical observations of the outputs from the actual process, $o = \{o_1, o_2, \ldots, o_N\}$, taken at times $\psi = \{\psi_1, \psi_2, \ldots, \psi_N\}$, the parameter calibration algorithm can be stated as: Solve

$$\underset{\{x_m\}, \{p_j\}}{\arg\min} \sum_{i=1}^{N} \|o_i - \tilde{g}(x_{\psi_i}, u_{\psi_i})\|_2^2$$

where $\{p_j\}$ is the set of calibrated parameters embedded within the functions f and g; and subject to an equality constraint on all pairs of adjacent states, given as

$$(x_{k+1} - x_k) = hf(\hat{x}_{k+1/2}, \hat{a}_{k+1/2})$$

where

$$\hat{x}_{k+1/2} = \frac{(x_{k+1} + x_k)}{2}$$

and

$$\hat{a}_{k+1/2} = L_{11}\tilde{g}(\hat{x}_{k+1/2}, u_k) + L_{12}u_k$$

Hence, a constrained minimization is specified, wherein the given equality constraints require that the choices for consecutive state vectors must be compatible with an approximate numerical integration of the state evolution equations as performed by the Midpoint Rule. In general, the minimization seeks to eliminate the error between the predicted output and the observed output at the points where the observations were made. It is likely that certain elements of the output are either more important, or are better indicators of a correct calibration, hence the 2-norm will likely be replaced in practice by a weighted vector norm, e.g., $\|a\|_Q^2 = a^T Q a$ for some positive definite Q.

5.2. Calibrating with Interval Scale Data

Measurement of the state of software development processes is notoriously difficult. It is often far easier to characterize the *differences* in the state over time. For example, Ref. 9 notes that one has no mechanism to know the number of undetected defects remaining in a test target, but one can certainly collect data about how many defects are detected per unit time. In Ref. 9 the authors develop and validate a calibration technique that refines a guess at the internal state given successive measurements of the observable changes to that hidden state.

By treating measurement of the actual process as interval scale data, we restrict our consideration to only differences between observations. Because there is no meaning assigned to zero on an interval scale, the calibration data can be translated by an arbitrary error vector with no consequence. That is, a calibration mechanism built to use interval scale data is tolerant to uniform constant measurement error.

Given the adapted Improved Euler numerical differential equation solver above, and recalling that for the modeling components described in Section 2, the general subsystem outputs, b, can be explicitly computed given only the system state vector, x, and the external input vector, u (Section 4.1), one can state that there exists an explicit function $b = \tilde{g}(x, u)$. In light of this observation, the discrete simulation step formula can be re-written as a function of the state and external inputs by substitution as follows:

$$m_t(x_t, u_t) = f(x_t, L_{11}\tilde{g}(x_t, u_t) + L_{12}u_t)$$

$$\hat{x}_{t+h}(x_t, u_t) = x_t + hm_t(x_t, u_t)$$

$$a_{t+h}(x_t, u_t, u_{t+h}) = L_{11}\tilde{g}\left(\hat{x}_{t+h}(x_t, u_t), u_{t+h}\right) + L_{12}u_{t+h}$$

$$m_{t+h}(x_t, u_t, u_{t+h}) = f(\hat{x}_{t+h}(x_t, u_t), a_{t+h}(x_t, u_t, u_{t+h}))$$

$$\bar{x}_{t+h}(x_t, u_t, u_{t+h}) = x_t + \frac{h}{2}\left(m_t(x_t, u_t) + m_{t+h}(x_t, u_t, u_{t+h})\right)$$

It is apparent that this step function can be composed recursively into a ϕ-step state-update function of the form

$$\bar{x}_{t+\phi h}(x_t, u_t, u_{t+h}, u_{t+2h}, \ldots, u_{t+\phi h})$$

Note that this multi-step state update function can be interpreted as an approximate discretization of the model equations using the adapted Improved Euler method of Section 4.1 as a numerical quadrature for integrating the non-linear state-derivative equations over each time-step under a Zero-Order Hold (ZOH) constraint on the external inputs, u.

For any pair of model states separated by ϕ time-steps, as generated by the adapted Improved Euler solution of the initial value problem specified by the initial state, x_{t_0}, the ZOH-constrained external inputs $\{u_i\}$ for $i \in \{t_0 + kh : k = 0, 1, \ldots\}$, and the model equations, the following function gives their difference:

$$d_\phi(x_{t_0}, u_{t_0}, u_{t_0+h}, \ldots, u_{t_0+\phi h}) := \bar{x}_{t+\phi h}(x_{t_0}, u_{t_0}, u_{t_0+h}, \ldots, u_{t_0+\phi h}) - x_{t_0}$$

This function can then be converted to a function of *output difference* rather than the *state difference*:

$$\tilde{d}_\phi(x_{t_0}, u_{t_0}, u_{t_0+h}, \ldots, u_{t_0+\phi h}) := \tilde{g}(\bar{x}_{t+\phi h}(x_{t_0}, u_{t_0}, u_{t_0+h}, \ldots, u_{t_0+\phi h}),$$
$$u_{t_0+\phi h}) - \tilde{g}(x_{t_0}, u_{t_0})$$

Given a set of interval scale measurements taken on the output of the actual process, one now has a mechanism for scoring a guess at the parameters and internal state of the model: First, one identifies the earliest measurement and notes its time as t_S. By taking the difference of all pairs (the approach of using all pairs rather than just adjacent pairs is developed and evaluated in Ref. 26) of observations, one generates a table of 3-tuples, where the columns are: (i) offset (in time) from t_S to the earliest observation in the pair; (ii) the magnitude of the offset in time between the two observations in the pair; and (iii) the difference in the observations taken by subtracting the earlier observation from the later. That is,

Time Offset from t_S to Pair (θ)	Duration Spanned (σ)	Output Delta (δ)
(integer multiples of h)	(integer multiples of h)	(Real vector)

To calibrate the model parameters and estimate the state of the model at time t_S in order to best match the interval scale empirical data captured from the actual process, one solves the unconstrained minimization:

$$\underset{x_{t_S}, \{p_j\}}{\arg\min} \sum_{i=1}^{N} \| \delta_i - \tilde{d}_{\sigma_i}$$
$$\times (x_{t_S} + d_{\theta_i}(x_{t_S}, u_{t_S}, \ldots, u_{t_S+\theta_i}), u_{t_S+\theta_i}, \ldots, u_{t_S+\theta_i+\sigma_i}) \|_2^2$$

where $\{p_j\}$ is the set of model parameters to be calibrated, and N is the number of entries in the table of calibration data. Again, it is likely that certain elements of the output are either more important, or are better indicators of a correct calibration, hence the 2-norm will likely be replaced in practice by a weighted vector norm, e.g., $\|a\|_Q^2 = a^T Q a$ for some positive definite Q.

As above, empirical observations are assumed to be spaced sparsely over a regular time grid, $G = \{t_0 + kh : k = 0, 1, \ldots\}$ — note as a consequence, that the empirical measurements do not necessarily capture the external inputs at the time-step granularity required by the functions $d(\cdot)$ and $\tilde{d}(\cdot)$. We therefore assume that the external inputs are piecewise constant, and that the grid points at which they are updated are recorded, hence the external inputs present at any grid point can be reconstructed. As a natural consequence, a zero-order hold constraint on the external inputs is satisfied.

The authors note that the choice of constructing the state difference function from the adapted Improved Euler method versus, say, the Midpoint Rule, was arbitrary. Intuitively, however, it seems that the parameters generated under this choice are more likely to produce a simulation that matches the empirical observations when using the adapted Improved Euler method to solve the initial value problem that defines the simulation. Also, the choice to embed the state estimation directly into the minimization problem rather than performing a constrained minimization (as was done for the other calibration routine), was made to evaluate the option. Depending on the minimization algorithm, one option may present more amenable properties. For example, if the minimization technique does not support equality constraints, one would embed the equations. Alternatively, if the minimization routine builds a numerical approximation of the Hessian matrix of the function to be minimized, then it will be a better choice to expose the state vectors as explicit arguments in the minimization in order to reduce the complexity of the function being minimized (and to improve the chance that it is convex in the arguments over which it is minimized).

6. Discussion

The preceding model exposition and simulation has demonstrated that the complex behavior of schedule-constrained software development activities can be captured through the formalism of state modeling. Furthermore, the simulation process for such a model is relatively simple, opening the way for iterative as well as deterministic decision-support applications.

For example, Section 4.2.1 notes that an idle workforce is recognizable as a discrepancy between the productive capability of the workforce, and the out-flow rate of the associated queue. One can capture this with a performance index; e.g.,

$$J(x, \chi, u, \Delta u) = \frac{1}{2} \sum_{a \in A} \left(q^a \int_{t_0}^{t_f} \|c_t^a - r_t^a\|_2^2 + \|u_t^a + \Delta u_t^a\|_R^2 \, dt \right) + \|\chi - x_{t_f}\|_Q^2$$

where A is the set of development activities, q^a is the weight given to activity a to specify the cost of its workforce members sitting idle, $c_t^a \in x$ is the productive capability of the team for activity a at time t, $r_t^a \in x$

is the out-flow rate of the queue of activity a at time t, $u_t^a \in u$ is the vector of externally controlled inputs for activity a at time t, $\Delta u_t^a \in \Delta u$ is the set of control inputs supplied to activity a at time t, and R is positive definite matrix assigning relative cost to the element of the input vector. For the process duration spanning $[t_0 \quad t_f]$, $x_{t_f} \in x$ gives the final state of the process. χ gives the desired final state of the process, and Q is a positive definite matrix assigning the cost of deviation between the actual final state and the expected final state (Q should be sufficiently large that the cost of allocating workforce to the process is much smaller than the cost of failing to meet the deadline). Given such a setup, one can quantitatively evaluate alternatives for eliminating idle workforce scenarios. For example, one could take a step toward optimal workforce allocation by solving

$$\underset{x, \Delta u}{\arg\min} \ J(x, \chi, u, \Delta u)$$

subject to the initial state, x_0, and the model equations. This yields the optimal control inputs, Δu_t, that best allocate and reallocate the minimum number of workforce members required to complete the project by the deadline. Note that this specification of the performance index ignores many practical problems, such as the fact that the minimization may choose to reallocate the workforce continually, or that there are limits on the process quality control parameter. Hence additional constraints would be necessary to successfully perform decision support.

The ultimate goal of this work is to develop the capability to optimize software development processes through the techniques of modern control theory. By developing a framework for building general software process models within the formalism of State Modeling, this work has taken a step toward that end. We hope to open the door to interesting control approaches for the various decision support needs in software project management, though we suspect that the modeling framework we have proposed herein is sufficiently general to apply outside the domain of software development (indeed, there are no aspects of the modeling work that are specific to software development).

References

1. Function Point Counting Practices Manual, International Function Point Users Group (IFPUG), Release 4.0, 1994.

2. Abdel-Hamid, T., Madnick, S.E.: *Software Project Dynamics*, Prentice Hall, 1991.
3. Boehm, B.W.: *Software Cost Estimation with COCOMO II*, Prentice Hall, 2000.
4. Cai, K.Y., Cangussu, J.W., DeCarlo, R.A., Mathur, A.P.: An overview of software cybernetics, *Proceedings of the Eleventh Annual International Workshop on Software Technology and Engineering Practice*, September 19–21, 2003, pp. 77–86.
5. Cai, K.-Y., Jing, T., Bai, C.-G.: *Partition Testing with Dynamic Partitioning*, COMPSAC, Proceeding of, Edinburgh, Scotland, July 2005.
6. Camacho, E.F., Bordons, C.: *Model Predictive Control*, Springer Publications, January 31, 2004.
7. Cangussu, J.W., Decarlo, R.A., Mathur, A.P.: A formal model for the software test process, IEEE Transaction on Software Engineering, 2002, Number 28, pp. 782–796.
8. João W. Cangussu, Kai-Yuan Cai, Scott D. Miller, Aditya P. Mathur: Software Cybernetics, Standard Article in *Encyclopedia of Computer Science and Engineering*, John Wiley and Sons, December 2007.
9. Cangussu, J.W., Karcich, R.M., Mathur, A.P., DeCarlo, R.A.: *Software Release Control Using Defect-Based Quality Estimation*, ISSRE, Proceedings of, pp. 440–450, November 2004.
10. Card, D.: *Statistical Process Control for Software*, IEEE Software, 11, pp. 1995–1997.
11. Chulani, S., Boehm, B.: Modeling software defect introduction and removal: COQUALMO (COnstructive QUAlity MOdel), Jan 1999.
12. Csíkszentmihályi, M.: *Flow: The Psychology of Optimal Experience*, Harper and Row, New York, 1990.
13. DeCarlo, R.A.: The component connection model for interconnected systems: Philosophy, problem formulations, and solutions, *Large Scale Systems: Theory and Applications*, 7, 1984, pp. 123-138.
14. DeCarlo, R.A., Saeks, R.: *Interconnected Dynamical Systems*, New York: Marcel-Dekker, 1981.
15. DeCarlo, R.A.: *Linear Systems: A State Variable Approach with Numerical Implementation*, Upper Saddle River, New York: Prentice Hall, 1989.
16. De Schutter, B., van den Boom, T.: Model predictive control for discrete-event and hybrid systems. Workshop on Nonlinear Predictive Control (Workshop S-5) at the 42nd IEEE Conference on Decision and Control, Maui, Hawaii, Dec. 2003.
17. Donzelli, P.: Decision support system for software project management, *IEEE Software*, 23(4), 2006, pp. 67–75.
18. Forrester, Jay W.: Industrial dynamics. Pegasus communications, 1961.
19. Marakas, G.M., *Decision Support Systems in the 21st Century*, Prentice Hall, 2003.
20. Martin, R.H., Raffo, D.M.: A model of the software development process using both continuous and discrete models, *International Journal of Software Process Improvement and Practice*, 5, 2000, pp. 147–157.

21. Martin, R., Raffo, D.: *Application of a Hybrid Process Simulation Model to a Software Development Project*, PROCEED, Proceedings of, 2006, pp. 237–246.

22. Miller, S.D., DeCarlo, R.A., Mathur, A.P.: Modeling and control of the incremental software test process, In *Proceedings of the 28th Annual International Computer Software and Applications Conference*, COMPSAC 2004, Vol. 2, 2004, pp. 156–159.

23. Miller, S.D., Mathur, A.P., DeCarlo, R.A.: DIG: A tool for software process data extraction and grooming, In *Proceedings of the 29th Annual International Computer Software and Applications Conference*, COMPSAC 2005, Vol. 1, July 26–28, 2005, pp. 35 – 40.

24. Miller, S.D., DeCarlo, R.A., Mathur, A.P.: A software cybernetic approach to control of the software system test phase, In *Proceedings of the 29th Annual International Computer Software and Applications Conference*, COMPSAC 2005, Vol. 2, July 26–28, 2005, pp. 103–108.

25. Miller, S.D., DeCarlo, R.A., Mathur, A.P.: A control-theoretic aid to managing the construction phase in incremental software development, In *Proceedings of the 30th Annual International Computer Software and Applications Conference*, COMPSAC 2006, Vol. 2, September 17–21, 2006, pp. 341–343.

26. Miller, S.D., DeCarlo, R.A., Mathur, A.P., Cangussu, J.W.: A control-theoretic approach to the management of the software system test phase. Special section on Software Cybernetics, *Journal of Systems and Software*, (79)11, 2006, pp. 1486–1503.

27. Miller, S.D., DeCarlo, R.A., Mathur, A.P.: A quantitative modeling for incremental software process control, In *Proceedings of the 32nd Annual International Computer Software and Applications Conference*, COMPSAC 2008, July 28–August 1, 2008, pp. 830–835.

28. Park, R.: Software size measurement: A framework for counting source statements, CMU-SEI-92-TR-20, 1992, Software Engineering Institute, Pittsburg, PA.

29. Scacchi, W.: Process models in software engineering, in Marciniak, J.J. (ed.), *Encyclopedia of Software Engineering, 2nd Edition*, John Wiley and Sons, Inc, New York, December 2001.

30. Shewhart, W.A.: *Statistical Method from the Viewpoint of Quality Control*, Dover Publication, 1986.

31. Volterra, V.: Variations and fluctuations of the number of individuals in animal species living together. In *Animal Ecology*. McGraw-Hill, 1931. Translated from 1928 edition by R. N. Chapman.

Appendix A: Modeling the Example Process

Figure A.1 refines Fig. 2 in light of the preceding mathematical development. A typical development activity model requires six inputs, and generates five outputs. Notice the facility for subtracting the amount of rework from the cumulative out-flow before feeding back to the *dependent cumulative out-flow* input of the controller. This mechanism allows re-work work items to flow into the queue without having them influence the enforcement of the coordination constraint. In Table A.1 below, which details the interconnections used in the simulation example, such subtraction of re-work from the dependent cumulative out-flow can be seen in lines 17 and 18.

To reduce clutter and improve readability, the model components are given short names. Component names are constructed by the following scheme,

$$\langle type \rangle \langle id \rangle$$

where *type* is one of

    P = Workforce Productive Capability Model
    C = Project Schedule Controller

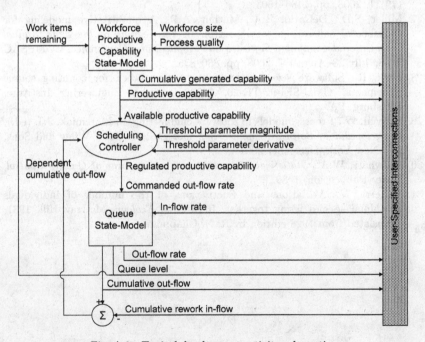

Fig. A.1   Typical development activity schematic.

Table A.1.   Interconnections made for the example simulation.

	Destination		Source		
	Component	Input	Component	Output	Scale
Interconnections feeding Feature Coding inputs					
*1	PFC	Workload Size	QFC	Queue Level	1
*2	CFC	Dependent Cumulative Out-Flow	QFC	Cumulative Out-Flow	1
*3	CFC	Available Productive Capability	PFC	Productive Capability	1
*4	QFC	Commanded Out-Flow Rate	CFC	Regulated Productive Capability	1
Interconnections feeding Test Case Coding inputs					
*5	PTC	Workload Size	QTC	Queue Level	1
*6	CTC	Dependent Cumulative Out-Flow	QTC	Cumulative Out-Flow	1
*7	CTC	Available Productive Capability	PTC	Productive Capability	1
*8	QTC	Commanded Out-Flow Rate	CTC	Regulated Productive Capability	1
9	QTC	In-flow Rate	QFAF	Out-Flow Rate	1
Interconnections feeding Test Case Execution inputs					
*10	PTE	Workload Size	QTE	Queue Level	1
*11	CTE	Dependent Cumulative Out-Flow	QTE	Cumulative Out-Flow	1
*12	CTE	Available Productive Capability	PTE	Productive Capability	1
*13	QTE	Commanded Out-Flow Rate	CTE	Regulated Productive Capability	1
14	QTE	In-Flow Rate	QTC	Out-Flow Rate	1
15	QTE	In-Flow Rate	QFCR	Out-Flow Rate	1
16	QTE	In-Flow Rate	QTCR	Out-Flow Rate	1
17	CTE	Dependent Cumulative Out-Flow	QFCR	Cumulative Out-Flow	−1
18	CTE	Dependent Cumulative Out-Flow	QTCR	Cumulative Out-Flow	−1
19	CTE	Threshold Parameter Magnitude	QFC	Cumulative Out-Flow	1
20	CTE	Threshold Parameter Derivative	QFC	Out-Flow Rate	1
Interconnections feeding Regression Testing inputs					
*21	PRT	Workload Size	QRT	Queue Level	1
*22	CRT	Dependent Cumulative Out-Flow	QRT	Cumulative Out-Flow	1

(*Continued*)

Table A.1.   (*Continued*)

	Destination		Source		
	Component	Input	Component	Output	Scale
*23	CRT	Available Productive Capability	PRT	Productive Capability	1
*24	QRT	Commanded Out-Flow Rate	CRT	Regulated Productive Capability	1
25	CRT	Threshold Parameter Magnitude	QFC	Cumulative Out-Flow	1
26	CRT	Threshold Parameter Derivative	QFC	Out-Flow Rate	1
Interconnections feeding Failure Analysis inputs					
27	PFA	Workload Size	QFAF	Queue Level	1
28	PFA	Workload Size	QFAT	Queue Level	1
29	CFA	Dependent Cumulative Out-Flow	QFAF	Cumulative Out-Flow	1
30	CFA	Dependent Cumulative Out-Flow	QFAT	Cumulative Out-Flow	1
*31	CFA	Available Productive Capability	PFA	Productive Capability	1
32	QFAF	In-Flow Rate	ADD	Feature Defect Detection Rate	1
33	QFAT	In-Flow Rate	ADD	Test Case Defect Detection Rate	1
34	QFAF	Commanded Out-Flow Rate	ASPL	O1	1
35	QFAT	Commanded Out-Flow Rate	ASPL	O2	1
Interconnections feeding Feature Correction inputs					
*36	PFCR	Workload Size	QFCR	Queue Level	1
*37	CFCR	Dependent Cumulative Out-Flow	QFCR	Cumulative Out-Flow	1
*38	CFCR	Available Productive Capability	PFCR	Productive Capability	1
*39	QFCR	Commanded Out-Flow Rate	CFCR	Regulated Productive Capability	1
40	QFCR	In-Flow Rate	QFAF	Out-Flow Rate	1
Interconnections feeding Test Case Correction inputs					
*41	PTCR	Workload Size	QTCR	Queue Level	1
*42	CTCR	Dependent Cumulative Out-Flow	QTCR	Cumulative Out-Flow	1
*43	CTCR	Available Productive Capability	PTCR	Productive Capability	1

(*Continued*)

Table A.1. (*Continued*)

	Destination			Source		
	Component	Input	Component	Output	Scale
*44	QTCR	Commanded Out-Flow Rate	CTCR	Regulated Productive Capability	1
45	QTCR	In-Flow Rate	QFAT	Out-Flow Rate	1
Interconnections feeding Defect Estimation inputs					
46	ADE	Features Produced	QFC	Cumulative Out-Flow	1
47	ADE	Feature Defects Removed	QFCR	Cumulative Out-Flow	1
48	ADE	Test Cases Produced	QTC	Cumulative Out-Flow	1
49	ADE·	Test Case Defects Removed	QTCR	Cumulative Out-Flow	1
Interconnections feeding Defect Detection inputs					
50	ADD	Estimated Feature Defects	ADE	Feature Defects Present	1
51	ADD	Regression Test Execution Rate	QTE	Out-Flow Rate	1
52	ADD	Estimated Test Case Defects	ADE	Test Case Defects Present	1
53	ADD	New Test Execution Rate	QTE	Out-Flow Rate	1
Interconnections feeding Proportional Splitter inputs					
54	ASPL	Reference R1	QFAF	Queue Level	1
55	ASPL	Reference R2	QFAT	Queue Level	·1
56	ASPL	Source	CFA	Regulated Productive Capability	1

Q = Queue Model

A = Other Algebraic Model

and *id* is from

FC = Feature Coding

TC = Test Case Coding

TE = New Test Case Execution

RT = Regression Test Case Execution

FA = Failure Analysis (Suffixed with T or F to indicate the individual queues.)

FCR = Feature Correction

TCR = Test Case Correction

DE = Defect Estimation Model

DD = Defect Detection Model

SPL = Proportional Splitting Component

Table A.1 gives a description of each interconnection by specifying the source and destination of each interconnection, and the scale factor that was used when specifying the connection. As mentioned in Section 2.4, the simulation was constructed with scale factors of unit magnitude (though sometimes negative) in order to focus on the model construction and behavior. In Table A.1, entries with an asterisk in the left-most column are inherent in the composition of the three modeling elements (productivity model, queue, and scheduling controller) into a standard development activity model; other entries define flows between activities. Entries in the table are grouped according to the development activity on the receiving end of the interconnection. Lastly, lines in the table are most easily read from right to left. For example, line 1 reads: "A direct connection (scale = 1) feeds the *queue level* element of the output vector of the feature coding queue into the *workload size* element of the input vector to the feature coding productive capability model."

Because this table specifies the entries of a sparse matrix as per Fig. 3, multiple interconnections to a single destination are summed to provide the total value of the destination element. There are 56 connections defined to implement the model shown in Fig. 1; a brief description of the interesting features of Table A.1 is given below.

Line 9 above specifies that failures traced to feature defects will, with some frequency, generate new test case specifications. This is shown as workflow 9 in Fig. 1. Line 14 specifies that the new test case coding activity passes its completed test cases to the new test case execution activity (workflow 3), and lines 15–16 specify that test cases must be re-executed upon correcting a defect (workflows 12 and 13). As mentioned above, lines 17 and 18 subtract this test re-execution re-work from the cumulative out-flow before feeding it as the dependent cumulative out-flow input to the scheduling controller. Lines 19 and 20 configure the new test case execution scheduling controller to regulate test execution as a function of the progress of the feature coding activity (exactly what function of the feature coding progress will depend on the parameters given in Appendix B.) Lines 25 and 26 configure the regression test execution activity's scheduling controller for regulation as a function of feature coding progress as well. Recalling that the example scenario consists of two increments, with specific test cases for each, it is clear that the scheduling controllers are being configured to enforce the incremental nature of the example scenario.

At line 27, the construction of the failure analysis activity begins, as shown in Fig. 10. Lines 27 and 28 add the queue levels of the two failure analysis queues to provide the "blinded" workload size that the actual development process will see (as opposed to a priori visibility into which failure reports correspond to feature defects vs. test case defects.) Lines 29 and 30 sum the cumulative out-flow to give the proper blinded view of cumulative progress to the scheduling controller. Lines 32 and 33 connect the in-flow rates of the failure analysis queues to the corresponding defect detection rate outputs of the defect detection component. This is represented by workflows 5 and 6 merging into workflow 7 in Fig. 1. Lines 34 and 35 connect the commanded out-flow rate inputs of the failure analysis queues to the outputs of the proportional splitting component, which is part of the implementation of workflows 8, 10, and 11; line 40 completes the specification of workflow 10, and line 45 completes the specification of workflow 11.

Lines 46–53 are easily readable, and connect the defect estimation and defect detection models to the appropriate sources of data. In Section 2.9, it was mentioned that the defect estimation component could be embedded into the defect detection component; this approach has not been taken here. Lastly, lines 54 and 55 connect the failure analysis queue levels to the reference inputs on the proportional splitter, and line 56 connects the regulated productive capability from the failure analysis scheduling controller to the source input of the proportional splitting component.

Appendix B: Simulation Study Parameters

In setting up the example simulation, all of the queues use the time constant, $\tau = 7/100$. The productive capability model parameters are given in Table A.2.

Table A.2. Productivity model parameters for the example simulation.

Activity	α	β	F^{cap}	ξ
Feature Coding	−0.83	−0.043	4	5.6
Test Case Coding	−0.83	−0.02	4	1.6
New Test Case Execution	−0.83	−0.02	1	0.26
Regression Test Case Execution	−0.83	−0.02	1	0.36
Failure Analysis	−0.83	−0.02	1	1.87
Feature Correction	−0.83	−0.02	1	1.6
Test Case Correction	−0.83	−0.02	1	1.6

Except for the new test case and regression test case execution activities, all scheduling controller components are configured with a constant threshold function of 10,000 work items (i.e., practically no threshold). The threshold function configured for both the new test case execution activity and the refression test case execution activity is

$$c(x) = \begin{cases} 0 & \text{if } 0 \leq x < 9 \\ -140x^3 + 210x^2 & \text{if } 9 \leq x < 10 \\ 70 & \text{if } 10 \leq x < 29 \\ -2260x^3 + 3390x^2 + 70 & \text{if } 29 \leq x < 30 \\ 1200 & \text{if } 30 \leq x \end{cases}$$

This function is a step function whose transitions have been smoothed with cubic splines. The steps rise to levels at 70 and 1200 work items at approximately $x = 10$, and $x = 30$, respectively. The values of x at which the level transitions occur corresponds to the number of work items within the two feature coding increments in the example scenario. The function is trivially (and continuously) differentiable.

The defect introduction model and defect detection model have parameters as well, as given in Table A.3.

The initial states for the state models are given in Table A.4.

Lastly, the external inputs were held constant at the values in Table A.5.

Table A.3. Parameters for the defect introduction and detection components.

μ_f	μ_t	b_{fc}	b_{tc}	$A_{fc} = A_{tc}$	D_{fc}	D_{tc}
0.01	0.001	1.12	1	1	0.15	0.125

Table A.4. Initial state vectors for the state-based components in the example simulation.

Development Activity	Initial Queue State	Initial Productive Capability Model State
Feature Coding	$[0 \quad 30 \quad 0]^T$	$[0 \quad 0]^T$
Test Case Coding	$[0 \quad 200 \quad 0]^T$	$[0 \quad 0]^T$
New Test Case Execution	$[0 \quad 0 \quad 0]^T$	$[0 \quad 0]^T$
Regression Test Case Execution	$[0 \quad 200 \quad 0]^T$	$[0 \quad 0]^T$
Failure Analysis (F)	$[0 \quad 0 \quad 0]^T$	$[0 \quad 0]^T$
(T)	$[0 \quad 0 \quad 0]^T$	
Feature Correction	$[0 \quad 0 \quad 0]^T$	$[0 \quad 0]^T$
Test Case Correction	$[0 \quad 0 \quad 0]^T$	$[0 \quad 0]^T$

Table   A.5.   Constant   control   inputs supplied during the example simulation.

Development Activity	ω_t	γ_t
Feature Coding	5	0.8
Test Case Coding	5	0.8
New Test Case Execution	5	0.8
Regression Test Case Execution	5	0.8
Failure Analysis	5	0.8
Feature Correction	2	0.8
Test Case Correction	2	0.8

Chapter 8

PROACTIVE MONITORING AND CONTROL OF WORKFLOW EXECUTION IN ADAPTIVE SERVICE-BASED SYSTEMS

STEPHEN S. YAU and DAZHI HUANG

School of Computing, Informatics,
and Decision Systems Engineering, Arizona State University,
Tempe, AZ 85287-8809, USA
{yau, dazhi.huang}@asu.edu

Systems based on service-oriented architecture are called *service-based systems* (*SBS*), and comprise of computing services offered by various organizations. Users of SBS often require these services to be composed into workflows to perform their high-level tasks. The users usually have certain expectations on the overall QoS (quality of service) of their workflows. Due to the highly dynamic environments of applications of SBS, in which temporary unavailability or quality degradation of services may occur frequently and unexpectedly, monitoring and controlling the execution of workflows adaptively in SBS are needed and should be done in distributed and proactive manner. In this chapter, important research issues and the current state-of-the-art will first be discussed. Then, a virtual machine-based architecture for the execution, monitoring and control of workflows in SBS, and a process calculus for modeling distributed monitoring and control modules are introduced. Using the virtual machine-based architecture and the process calculus, our approach to synthesizing software modules for proactive monitoring and control of workflow execution in SBS is presented.

1. Introduction

Recent development of service-oriented computing and grid computing has led to rapid adoption of service-oriented architecture (SOA) in distributed computing systems, such as enterprise computing infrastructures, grid-enabled applications, and global information systems. One of the most important advantages of SOA is the capability that enables rapid composition of the needed services provided by various service providers through networks for distributed applications. Software systems based on SOA are called *service-based systems* (*SBS*), and often comprise of computing services offered by various organizations. These

239

computing services provide well-defined interfaces for users to access certain capabilities offered by various providers, and are often hosted on geographically-dispersed computer systems. Due to different system capacities, active workloads or service contracts with users, such services usually provide various QoS, such as timeliness, throughput, accuracy, security, dependability, and survivability. Users of SBS will discover and access the most suitable services, which not only provide the required functionalities, but also meet the expected QoS of users.

Besides direct access to individual services, users often want to carry out workflows composing of various services in SBS to perform high-level tasks that cannot be done by an individual service. The composition of services can be automated based on the required functionality, referred as the *goal*, of the entire workflow.[1-3] However, the user of a workflow often has certain QoS expectations on the entire workflow, such as a deadline for completing certain tasks of the workflow.

A major problem for the development of high-quality SBS is how to satisfy users' multiple QoS requirements simultaneously in highly dynamic operating environments. Although certain QoS aspects in service composition can be incorporated,[4,5] the execution of workflows in SBS may not satisfy users' requirements due to the highly dynamic environments of SBS, in which temporary unavailability or quality-degradation of services may occur unexpectedly. Furthermore, because the satisfaction of requirements in an aspect of QoS often requires certain sacrifice in other QoS aspects, it is necessary to adaptively control the tradeoffs of requirements among multiple QoS aspects in SBS.

From software cybernetics perspective,[6,7] such a problem can be tackled by constructing a close-loop control-based SBS, which is capable of monitoring the workflow executions, and dynamically controlling the selection and configuration of services in the workflows to meet users' requirements. Such an SBS is referred as *an adaptive SBS (ASBS)*. The concept of ASBS is depicted in Fig. 1, in which functional services used to compose the ASBS and the modules for monitoring and controlling QoS form a closed control loop.[8] The QoS monitoring modules collect the measurements of various QoS aspects as well as system status concerning the QoS aspects used to determine the QoS adaptation for adjusting the configurations and service operations of ASBS to satisfy various QoS requirements simultaneously. However, the distributed and loosely-coupled nature of SBS has imposed many challenges for developing such an SBS. First, it requires proactive monitoring and control of workflow execution

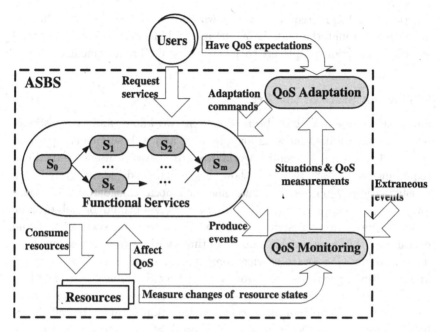

Fig. 1. A conceptual view of ASBS.

because various problems affecting workflow execution in ASBS need to be detected and workflows in ASBS need to be adapted accordingly in a timely manner. Proactive monitoring and control of workflow execution will greatly increase the chances of completing the workflows satisfactorily. For example, if a service to be used in a workflow fails, it will allow more time to find a replacement for the failed service and reconfigure the workflow if the failure can be detected before the failed service needs to be invoked.

Second, it requires distributed execution monitoring and control to overcome the inefficiency and potential security or dependability problems of centralized monitoring and control approaches.

In this chapter, we will first introduce a virtual machine-based architecture for executing, monitoring and controlling workflows in ASBS, and a process calculus for modeling distributed monitoring and control modules. Then, we will present our approach to synthesizing software modules for proactive monitoring and control of workflow execution in ASBS. In our approach, proactive distributed execution monitoring and control modules will be generated based on the specifications of both functional and QoS requirements of workflows, and will be coordinated in runtime through the underlying virtual machines to monitor and control workflow execution adaptively to satisfy users' QoS requirements.

2. Current State of the Art

Substantial research has been done in workflow and web service research communities on workflow planning and web service composition to construct and execute workflows to satisfy users' functional requirements.[1-3,9,10] Recently, how to design and adapt systems to satisfy various QoS requirements has also been considered.[11-15] QoS-aware service composition[11-13] aims at finding optimal or sub-optimal service composition satisfying various QoS constraints, such as cost and deadline, within a reasonable amount of time. Various techniques have been developed for QoS-aware service composition, such as service routing[11] and genetic algorithms.[12,13] However, the QoS models considered in existing QoS-aware service composition methods are usually very simple, and runtime adaptation of service composition cannot be efficiently handled by existing QoS-aware service composition methods. Self-tuning techniques[14] and autonomic middleware[15] were developed for providing support for service adaptation and configuration management. In Ref. 14, an online control approach was presented to automatically minimizing power consumption of CPU while providing satisfactory response time. In Ref. 15, an architecture for an autonomic computing environment, called AUTONOMIA, was presented, including various middleware services for fault detection, fault handling and performance optimization. However, these techniques and middleware are not designed for SBS, which is more loosely-coupled and difficult to control since SBS is often composed by multiple service providers.

Execution monitoring of workflows has been considered as an integral part of workflow management systems. Recently, grid workflow systems,[16-20] which operate in dynamic environments requiring execution monitoring and control similar to SBS, have been presented. DAGMan[16] provides a limited query capability for checking job status, which requires additional programming effort if some application-specific monitoring needs to be done. Gridbus[17] monitors workflow execution status based on event notification, and uses tuple spaces to exchange various event tuples indicating the execution status of distributed jobs. GridFlow[18] provides workflow execution and monitoring capabilities based on a hierarchical agent-based resource management system (ARMS)[21] for Grid, in which the agents serve as representatives of resources and are able to record performance data of the resources. Kepler[19] provides an easy-to-use environment for the design, execution and monitoring of scientific workflows. It relies upon specialized actors, which are interfaces to services on Grid, to provide job submission, execution and monitoring capabilities. Pegasus[20] has a Concrete Workflow Generator to automate the creation of a workflow in the form of DAG (Directed Acyclic Graph), and uses systems like DAGMan to execute and monitor workflows. However, the workflow monitoring capabilities in these Grid workflow systems as well as in many other workflow systems have a similar characteristic· They only focus on the status of the current workflow execution, and not proactively acquiring and analyzing status information related to future workflow execution.

3. Background

Our research on proactive monitoring and control of workflows in ASBS is based on α-Calculus[27,28] and our Workflow Virtual Machine (WVM).[30] The α-Calculus provides a high-level, platform-independent programming model for distributed software modules in SBS. WVM provides a platform-independent environment for executing, monitoring and controlling workflows in SBS. Hence, before describing our approach (see Section 4), we will first give an overview of the α-Calculus and WVM in this section.

3.1. *Workflow Virtual Machine*

Since a SBS often spans across heterogeneous networks consisting of various platforms, it is difficult for developers to handle the differences in hardware and software in these platforms. A virtual machine can mask such differences among various platforms, and provide a platform-independent environment for developing and executing application software. We have developed a virtual machine-based architecture to provide the essential functionalities for executing, monitoring and controlling workflows in SBS.[30] Figure 2 shows our virtual machine-based architecture, called workflow virtual machine (WVM). In WVM, workflow (WF) agents, monitors, and controllers are responsible for executing, monitoring, and controlling the adaptation of workflows in SBS, respectively.

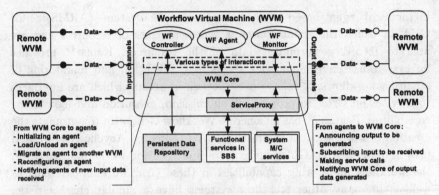

Fig. 2. Our virtual machine-based architecture for executing, monitoring and controlling workflows in SBS.

Our WVM is an extension of the SINS (Secure Infrastructure for Networked Systems) virtual machine.[22] Each of our WVMs has a set of input/output channels, uniquely identified by their names, and communicates with each other through these named channels to exchange data for coordinating activities of agents running on the WVMs. The communication over these named channels is supported by SPREAD,[23] which is a reliable group communication package. Each agent running on a WVM will take certain input data received by the WVM, and send the generated data, if any, through the WVM. Some specific named channels are reserved for exchanging control messages generated by virtual machines, such as load/unload messages for an agent.

Like SINS, our WVM provides the capabilities for loading and unloading agents, accepting an agent's subscriptions on the input data required by the agent as well as announcements on the output data that will be generated by the agent, sending and receiving the data through the named channels, and notifying the agents when new input data is available. In addition, our WVM has the following capabilities:

- Parametrized initialization of agents, which subsequently enables runtime reconfiguration of agents when workflows need to be adapted due to user requirements or environment changes.
- Invocation of functional services in SBS and system monitoring and control (M/C) services, such as system performance monitoring packages, network management and migration, to support workflow execution, monitoring and control in SBS.

- Event logging into a persistent data repository for recording workflow execution history.

The WVM Core shown in Fig. 2 is mainly responsible for these capabilities. The ServiceProxy in the WVM provides an interface for each functional service or system M/C service in SBS. When an agent (a WF agent, WF monitor or WF controller) evaluates its input data and makes a decision to invoke a certain service, the WVM handles the service invocation as follows:

i. The agent generates a service invocation request with the necessary information, including the name of the service to be invoked, the identity of the user invoking this service if access control for the service is required, and the name of the input channel, where the agent expects to receive the result of this service invocation.

ii. The agent puts the request into an output channel named *ServiceInvoke*.

iii. The WVM Core parses the service request, performs security checking, if necessary, and invokes the corresponding interface in the ServiceProxy.

iv. Once invoked, the corresponding interface in the ServiceProxy constructs an appropriate service request or API call, depending on whether the service to be invoked is a Web Service or other local system services. The constructed service request or API call will then be executed by the corresponding services.

v. When the ServiceProxy receives the result of the service invocation, it puts the result in the input channel specified by the agent in (i).

The above service invocation process in the WVM not only provides agents the capability to make asynchronous service invocations and perform other tasks in parallel, but also provides richer information regarding service invocations, which will be very useful in our monitoring approach for workflow execution. For example, service delay can be easily measured in the ServiceProxy to capture more semantic-rich events, such as unexpected delay due to network failure and insufficient computing resources or other causes not just the information on the success or failure of service invocations. Our virtual machine-based architecture provides the necessary capabilities for WF monitors to perform execution monitoring, subscribe to important events, such as the status of service invocations, receive events notifications from WVMs when such events occur, and invoke necessary system M/C services without knowing the low-level details of such services.

3.2. α-Calculus

The α-calculus is used in our approach to proactive monitoring and control of workflows in ASBS for modeling distributed monitoring and control modules. Process calculi have been used as programming models for concurrent and distributed systems.[24,25] We have developed the α-calculus, based on the classical process calculus,[26] to provide a formal programming model for SBS.[27,28] The α-calculus has well-defined operational semantics involving interactions of external actions and internal computations for assessing the current situation and responding to it for system adaptation.[27,28] The external actions include communication among processes, logging in and out of groups/domains. The internal computations include invocation of services as well as internal control flow.

For the sake of completeness, we summarize in Table 1 the part of the syntax of the α-calculus which will be used in this chapter. Similar to a classical process calculus, a system in the α-calculus can be the parallel composition of two other systems, or a recursive or non-recursive process. A recursive or non-recursive process can be an inactive process, a nominal identifying a process, a process performing external actions, a process performing internal computations, a service exporting a set of methods, for users to invoke or the parallel composition of two other processes. The methods are described by their preconditions and postconditions specifying the constraints on the inputs accepted and outputs provided by the methods, respectively. In Table 1, $I : l_i(y)$ denotes the invocation of the method l_i exported by a service I using parameter y. External actions involve the input and output actions on named channels with types as in the ambient calculus.[29] Internal computation involves beta reduction, conditional evaluation for logic control, and invocation of public methods exported by a named service or private methods exported by the process itself.

4. Synthesizing Software Modules for Proactive Monitoring and Control of Workflow Execution in ASBS

In this section, we will present our approach to synthesizing software modules for proactive monitoring and control of workflow execution in ASBS.

Our approach to synthesizing software modules for proactive monitoring and control of workflow execution in ASBS consists of the following three steps:

Table 1.   A partial syntax of the α-calculus to be used in our approach.

(System)			
$S ::=$		$N ::=$	
$fix\ I{=}P$	(recursive or non-recursive process)	x	(name variable)
$S\|S$	(parallel composition of two systems)	n	(name)

(Processes)		**(External actions)**	
$P ::=$		$E ::=$	
0	(inactive process)	K	(Communication actions)
$P\|P$	(parallel composition)		
I	(identifier)	$K ::=$	
$E.P$	(external action)	$Ch(x)$	(input)
$C.P$	(internal computation)	$Ch{<}Str{>}$	(output)
$P\{l_1(x_1),\ldots l_k(x_k);\ldots l_n(x_n)\}$	(method export)		
$l_1,\ldots l_k$ are private methods that can		$Ch ::=$	
be invoked by P itself only while			
$l_{k+1},\ldots l_n$ are public methods that can			
be invoked by other processes.			
		N	(named channel with type)

(Internal computations)			
$C ::=$		$pre ::= \sigma[y] \wedge \rho[y]$	
$let\ x = D\ instantiate\ P$	(beta reduction)		
$if\ C(x)\ then\ P\ else\ P'$	(conditional evaluation)	$post ::= (\sigma[x] \wedge \rho[x])x$	
ρ	(constraint)		
ε	(no-computation)	$\sigma ::=$	
tt	(constant true)	b	(base type)
$f\!f$	(constant false)	$\sigma \rightarrow \sigma$	(function type)
$D ::=$		$\rho ::=$	
$I : l_i(y)$	(method invocation)	$x \geq y + c$	
		$x > y + c$	
$I : l_i ::= pre_i :: post_i[y]$	(method definition)	$x \leq y + c$	
		$x < y + c$	

S-1) Construction of performance models for services to be used in the workflows in ASBS to support the estimation of QoS.

S-2) Synthesis of α-calculus descriptions of distributed processes for proactive monitoring and control of workflow execution.

S-3) Translation of process descriptions generated in *S*-2) to platform-specific code and compilation of the generated code to generate monitor and controller modules.

In this section, we will focus on **S-2)** of our approach. For **S-1)**, we have presented an approach to constructing service performance models in,[8,34] and will provide a brief overview of this modeling approach in Section 4.2. For **S-3)**, we have developed an α-calculus compiler to generate Java code,[27,28] and we are currently developing another α-calculus compiler to generate C# code. The rest of this section will be organized as follows: We will first discuss how WF agents, monitors, and controllers shown in Fig. 2 are coordinated to execute workflows, and monitor and control workflow execution in ASBS in Section 4.1. Then, we will present how to synthesize WF monitors and controllers in Sections 4.2 and 4.3, respectively.

4.1. *Workflow Execution, Monitoring and Control in ASBS*

• WF agents in ASBS

In Ref. 28, we have presented an approach to composing workflows and synthesizing WF agents from the composed workflows. Regardless of the complexity of the composed workflows, each synthesized WF agent has the internal control structure shown in Fig. 3.

A WF agent will first receive inputs, including service parameters and situation information, through named channels in WVM (see Section 3.1). Then, the WF agent will check certain conditions extracted from the synthesized workflow and invoke one or more services based on the conditions. The WF agent may receive additional inputs, check more conditions and invoke more services as illustrated by the dotted arrows and boxes in Fig. 3. Finally, the WF agent will send out results through named channels in WVM. For each composed workflow, one or more WF agents may be synthesized. These WF agents may invoke services in parallel if the composed workflow contains parallel sub-processes, or sequentially if a WF agent takes the results of other WF agents as inputs.

• Coordination among WF agents, monitors and controllers

To monitor and control the execution of a workflow in ASBS, we have defined a coordination protocol among WF agents, monitors and controllers for the workflow. This protocol consists of two phases: Phase 1) before service invocation, and Phase 2) during service invocation. In these two phases, the WF monitor M, the WF agent A, and the WF controller C

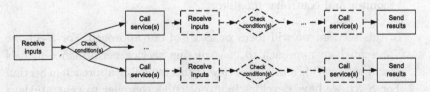

Fig. 3.   The internal control structure of a WF agent.

responsible for monitoring, invoking and controlling a service S used in the workflow interact with each other as follows:

Phase1) Before S is invoked, M, A, and C perform the following activities in parallel:

- **Activities of M in Phase1):** M executes the process shown in Fig. 4(a) periodically with a time interval t until S is invoked or a *TERMINATION* message from C is received. The process is described as follows:

  M-1.1) Collect the data related to the status of S as well as the status of the host H and network N, where S resides by invoking certain system monitoring services.

  M-1.2) If S, H or N becomes unavailable, notify all the related WF controllers to find replacements for the unavailable services used in the workflow. Then, go to *M-1.6)*. Otherwise, continue with *M-1.3)*.

  M-1.3) If the parameters for invoking S are not ready, go back to *M-1.1)*. Otherwise, continue with *M-1.4)*.

  M-1.4) Estimate the QoS that can be achieved based on the service parameters (see Ref. 8 and 34, and also the explanations in Section 4.2) and the current status of S, H and N, and compare the estimated QoS EQ with the QoS requirements QR for S in this workflow. If EQ does not satisfy QR, go to *M1.5)*. Otherwise, M sends out a *CHECKPOINT_CLEAR* message with a timeout t for S. Then, go back to *M1.1)*.

  M-1.5) M notifies C to reconfigure S or to find a replacement for S. Then, go to *M-1.6)*.

  M-1.6) M waits for a message from C. If the message is a SERVICE_REPLACEMENT message, M changes the monitoring target from S to the new service in the *SERVICE_REPLACEMENT* message. Go back to *M-1.1)*. If the message is a TERMINATION message, M terminates.

- **Activities of C in Phase1):** As shown in Fig. 4(b), C processes the notifications received from M and other WF controllers until S is invoked or a *TERMINATION* message from another WF controller is received. There are the following two cases:

  Case C-1.1) If S, H, or N becomes unavailable, C finds a replacement for S. If the replacement for S is found, C sends a *SERVICE_REPLACEMENT* message to M

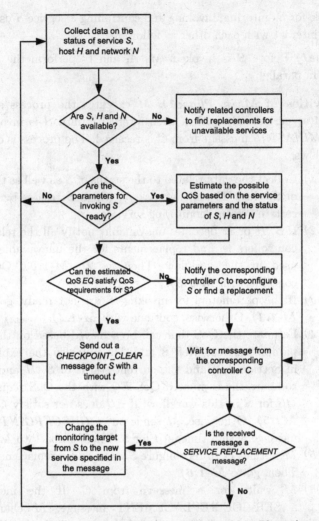

Fig. 4(a).   Process executed by a Monitor M in **Phase1**).

and A. If a replacement for S is not found, C sends a *TERMINATION* message to M and A as well as other WF controllers.

Case C-1.2) If EQ does not satisfy QR (or there is unexpected delay in the execution of a service invoked before S in the workflow), C finds a new service configuration for S, which makes the new EQ *to* satisfy QR (or reduce the expected delay for executing S).

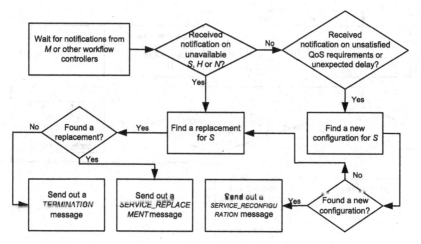

Fig. 4(b). Process executed by a Controller C in **Phase1**).

If a new service configuration is found, C sends a *SERVICE_RECONFIGURATION* message to M and A. If such a new service configuration is not found, C finds a replacement for S that can satisfy QR (or reduce the expected delay for executing S). If such a replacement is found, C sends a *SERVICE_REPLACEMENT* message to M and A. Otherwise, C sends out a *TERMINATION* message.

- **Activities of A in *Phase1*)**: As shown in Fig. 4(c), A waits for the parameters for invoking S, and processes control messages from M and C until S is invoked or a *TERMINATION* message from C is received. There are the following three cases:

Case A-1.1) If the parameters for invoking S are ready (either received from users or generated as the results of other services), A notifies M that the parameters are ready. Continue to wait for control messages from M and C.

Case A-1.2) If a *CHECKPOINT_CLEAR* message from M is received, A checks whether all the services that must finish before S starts have finished. If yes, A will invoke S immediately. Otherwise, A will wait until those services finish before invoking S. After invoking S, A notifies M and C that S is invoked, and enters the execution phase.

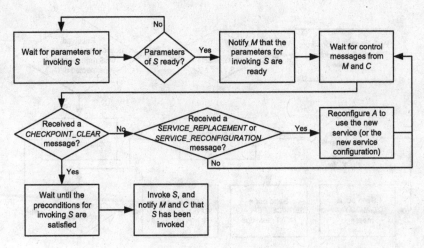

Fig. 4(c).   Process executed by a WF agent A in **Phase1)**.

Case A-1.3) If a *SERVICE_REPLACEMENT* or a *SERVICE_RECONFIGURATION* message from C is received, A will reconfigure itself to use the new service in the *SERVICE_REPLACEMENT* message or the new service configuration in the *SERVICE_RECONFIGURATION* message.

Phase2) M, A, and C perform the following activities in parallel during the execution of S:

- **Activities of M in Phase2)**: M executes the process shown in Fig. 5(a) periodically with a time interval t until the execution of S is complete or *TERMINATION* message from C is received:

  M-2.1) M collects the data related to the performance of S and the status of H and N.

  M-2.2) M records the elapsed time *delay* for executing S, and measures the current QoS CQ of S.

  M-2.3) If S, H or N becomes unavailable, M notifies all the related WF controllers to find replacements for the unavailable services. Then go to **M-2.8)**. Otherwise, continue with **M-2.4)**.

  M-2.4) If the *delay* is greater than *MAX_DELAY*, go to **M-2.6)**. Otherwise, continue with **M-2.5)**.

  M-2.5) If CQ satisfies QR, go to **M-2.1)**. Otherwise, go to **M-2.7)**.

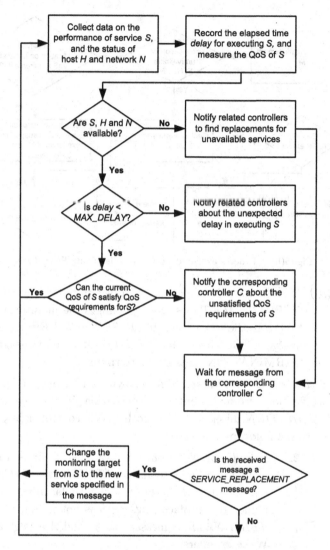

Fig. 5(a). Process executed by a monitor *M* in **Phase2)**.

M-2.6) *M* notifies the WF controllers of the unexpected delay in the execution of S. These WF controllers are responsible for the services to be invoked after *S* in the workflow. Then, go to **M-2.8)**.

M-2.7) *M* notifies *C* of the unsatisfied QoS requirements for *S*. Then, go to **M-2.8)**.

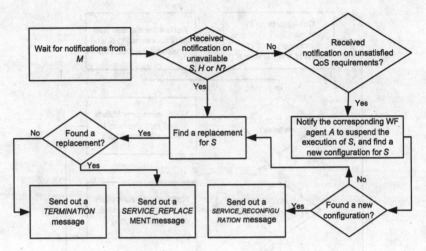

Fig. 5(b). Process executed by a controller C in **Phase2**).

M-2.8) M waits for a message from C. If the message is a *SERVICE_REPLACEMENT* message M changes the monitoring target from S to the new service in the *SERVICE_REPLACEMENT* message. Go back to **M-2.1)**. If the message is a TERMINATION message, M terminates.

• *Activities of C in Phase2)*: As shown in Fig. 5(b), C processes the notifications from M until the execution of S is complete or a *TERMINATION* message from another WF controller is received. There are the following two cases:

Case C-2.1) If S, H, or N becomes unavailable, C finds a replacement for S, and sends a *SERVICE_REPLACEMENT* message to M and A if the replacement for S is found. If a replacement for S is not found, C sends a *TERMINATION* message to M and A as well as other WF controllers.

Case C-2.2) If CQ does not satisfy QR, C first notifies A to suspend the execution of S. Then, C finds a new service configuration for S, which satisfies QR. If a new service configuration is found, C sends a *SERVICE_RECONFIGURATION* message to M and A. If such a new service configuration is not

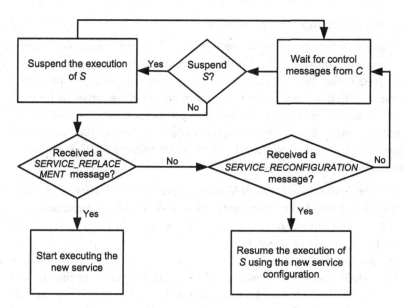

Fig. 5(c). Process executed by a WF agent *A* in **Phase2**).

found, *C* finds a replacement for *S* that satisfies *QR*. If such a replacement is found, *C* sends a *SERVICE_REPLACEMENT* message to *M* and *A*. Otherwise, *C* sends out a *TERMINATION* message.

- **Activities of *A* in Phase2):** As shown in Fig. 5(c), *A* waits for control messages from *C* until the execution of *S* is complete or a *TERMINATION* message from *C* is received. There are the following three cases:

Case A-2.1) If *C* notifies *A* to suspend the execution of *S*, *A* suspends the execution of *S* and waits for further control messages from *C*.

Case A-2.2) If a *SERVICE_REPLACEMENT* message from *C* is received, *A* reconfigures itself to use the new service in the message, and starts executing the new service.

Case A-2.3) If a *SERVICE_RECONFIGURATION* message from *C* is received, *A* reconfigures itself to use the new service configuration in the message, and resumes the execution of the reconfigured service.

4.2. *Synthesizing WF Monitors*

Our coordination protocol in Section 4.1 shows an important difference between our approach to workflow execution monitoring and those in Ref. 16–20: the WF monitors in our approach proactively monitor the services as well as computing and communication resources to be used in future workflow execution, so that the problems which may affect future workflow execution can be captured earlier. However, the following two issues need to be addressed for developing WF monitors with such proactive monitoring capability:

1. The proactive monitoring will have larger overhead since more services and resources need to be monitored at the same time. Such overhead, if not carefully handled, may negate the possible benefits in terms of the higher success rate of workflow execution.
2. A method to accurately estimate QoS before service invocations is needed. Without such a method, WF monitors cannot generate the *EQ* in *Phase1)* of our coordination protocol, and hence will not be able to notify WF controllers to adapt the workflow before invoking the services.

To address the first issue, we have presented an approach[30] to adaptive and proactive execution monitoring based on the philosophy similar to that of adaptive sampling designs of experiments,[31] in which experimental designs will be changed based on the data collected from previous experiments to improve the quality of experiments. The WF monitors synthesized using this approach will first monitor a subset of services and resources, and reconfigure themselves or load new monitors to acquire more information to create a clearer view of system status when any problem affecting WF QoS occurs. Many problems in networked computing systems have subtle logical connections among them. For example, a sudden slowdown of downloading a large file from a remote host may indicate a problem with the network connection to the remote host, or a problem with the remote host itself. Either problem may lead to failures for accessing other services and resources at the same host or using the same connection. Hence, for workflow execution monitoring, it is possible to select a subset of services and resources to be used in the workflow as the *indicators*, whose problems often lead to or co-occur with the problems of other services and resources. When problems with the indicators occur, WF monitors should proactively check the status of those services and resources related to the indicators. Hence, our approach in Ref. 30 can reduce the overhead of

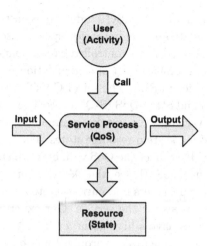

Fig. 6. The cause-effect chain of activity-state-QoS in SBS.

proactive execution monitoring while still being able to detect the problems occurred in workflow execution accurately and quickly.

In this subsection, we will focus on how to synthesize WF monitors to address the second issue.

- Performance models for estimating service QoS

In Ref. 8, we have presented an approach to constructing *Activity-State-QoS (ASQ)* models for the cause-effect dynamics in ASBS. Figure 6 depicts the cause-effect chain of user activities, system resource states, and QoS performance. A service request from a user calls for a system process, which will utilize certain resources and change the states of the resources in the system environment. The changes of the resource state in turn affect the QoS of the process. Hence, an ASQ model is a 6-tuple

$$\langle \mathcal{A}, S_0, S, Q, R_S, R_Q \rangle,$$

where $\mathcal{A}$ is a set of possible user activities that can be performed on a particular service, S_0 is a set of initial system resource states, S is the set of all possible system resource states, Q is the set of possible value for a particular aspect of QoS, R_S is a relation defined on $\mathcal{A} \times S_0 \rightarrow S$ representing how user activities and initial system resource states affect future system resource states, and R_Q is a relation defined on $\mathcal{A} \times S_0 \rightarrow Q$ representing how user activities and initial system resource states affect QoS aspects.

In our approach in Ref. 8, ASQ models are constructed through the analysis of the data related to user activities, system resource states and the QoS aspects, and collected from controlled experiments in SBS. The final result from the data analysis is a set of Classification And Regression Trees (CART) for each service in SBS. The set of CART trees are at two levels: Activity-State (A-S) and State-QoS (S-Q) levels. The A-S level CART trees reveal how activity variables (i.e. service parameters) drive the changes of state variables reflecting system resource states. The S-Q level CART trees reveal how state variables affect the QoS variables, which can be used as the measurement of service QoS. Hence, the QoS of a service can be estimated based on the service parameters and current system status using A-S level and S-Q level CART trees for the service. Our experimental results have shown that CART trees have a high accuracy for estimating QoS.[8]

The ASQ models generated through our data analysis can be incorporated in our WF monitors using the following two processes:

***P-1)* Automated transformation of a CART tree to the α-calculus terms in WF monitors.** Each CART tree represents a decision process for determining the value of the dependent variable based on the observed values of the independent variables. For example, Fig. 7 shows the graphical and text forms of the CART tree for the variable "Pool Paged Bytes" against the experimental control variables.

Each CART tree generated in our data analysis has a similar form as that shown in Fig. 7, and can be automatically transformed to an executable decision process embedded in our WF monitors using the following process:

T-1) Replace the variables in the tree with their names.
T-2) Store each line in the text form as a node in a tree data structure.

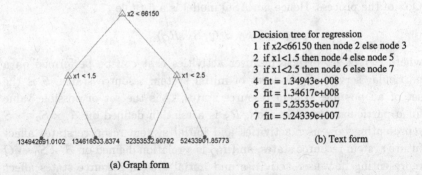

Decision tree for regression
1 if x2<66150 then node 2 else node 3
2 if x1<1.5 then node 4 else node 5
3 if x1<2.5 then node 6 else node 7
4 fit = 1.34943e+008
5 fit = 1.34617e+008
6 fit = 5.23535e+007
7 fit = 5.24339e+007

(b) Text form

(a) Graph form

Fig. 7. The CART tree for "Pool Paged Bytes" against the experimental control variables, where ×1 denotes the client number, and ×2 denotes the sampling rate.

T-3) Link the nodes based on their line numbers and the references to other lines in the text form.

T-4) Perform an in-order print of the tree data structure.

Following *T-1)–T-4)*, the CART tree in Fig. 7 can be transformed to the following decision process for estimating the value of "Pool Page Bytes" based on the client number and sampling rate:

```
0    tuple inputs<'sampling rate', 'client number'> //Retrieve inputs
1    if 'sampling rate' < 66150 then
2      if 'client number' < 1.5 then
4        tuple result<'Pool Paged Bytes', 1.34943e+008> //Output the
         estimation result
5      else tuple result<'Pool Paged Bytes', 1.34617e+008>
3    else if 'client number' < 2.5 then
6      tuple result<'Pool Paged Bytes', 5.23535e+007>
7    else tuple result<'Pool Paged Bytes', 5.24339e+007>
```

P-2) **Simplifying and merging multiple CART trees for efficient, adaptive execution monitoring.** For each service, multiple CART trees at A-S and S-Q levels will be generated for estimating the QoS aspects and system resource states. All these CART trees can be converted to executable decision processes as discussed in *P-1)*. However, if these decision processes are all embedded in our WF monitors separately, there will be large processing overhead because many branches in these CART trees correspond to normal operation conditions, in which the system resources are sufficient for the service to satisfy the QoS requirements of users. Furthermore, through the analysis of our experimental results, we have found that there is a small subset of independent variables used in all A-S (or S-Q) level CART trees for a service.

A process to simplify and merge multiple CART trees for a service, and generate executable code for more efficient and adaptive monitoring is presented as follows:

S-1) Follow *T-1)–T-3)* in *P-1)* to store all the CART trees for a service into tree data structures.

S-2) Traverse the leaf nodes of all the S-Q level CART trees to compare the estimated values of QoS counters with users' QoS requirements to identify all the leaf nodes corresponding to the conditions when the system does not satisfy users' QoS requirements.

S-3) Remove all the leaf nodes not identified in **S-2)** and all the non-leaf nodes not on any paths from the root node to the leaf nodes identified in **S-2)** to find the simplified S-Q level CART trees.

S-4) Extract the conditions associated with the remaining non-leaf nodes in each simplified S-Q level CART tree and consolidate the conditions into a list of 2-tuples $<var, range>$, where var is an independent variable (state counter) in at least one of the conditions, and $range$ is the minimum value range of this independent variable satisfying all the conditions related to this variable.

S-5) Merge the lists of 2-tuples $<var, range>$ generated in **S-4)** for all S-Q level CART trees to a list of 2-tuples $<var_M, range_M>$, such that a variable var_M only appears once in the merged list, and the $range_M$ of var_M is the minimum range covering all the ranges of var_M in all the lists containing var_M.

S-6) Generate decision processes following **T-4)** for the simplified S-Q level CART trees generated in **S-3)**.

S-7) For the merged list generated in **S-5)**.

S-7.1) Generate a conditional evaluation statement (if-statement) as follows:

> if $var_{M0} \in range_{M0} \wedge \ldots \wedge var_{Mn} \in range_{Mn}$ then
> Generate a "QoS requirement unsatisfied" event
> Call the dynamicLoadDecisionProcess procedure.
> else
> Wait to get the next observation
> where $<var_{M0}, range_{M0}>, \ldots,$ and $<var_{Mn}, range_{Mn}>$ are
> all the 2-tuples in the merged list.

S-7.2) Generate the *dynamicLoadDecisionProcess* procedure based on the lists generated in **S-4)** as follows:

For each list L generated in M4), generate the following conditional evaluation statement:

> if $var_0 \in range_0 \wedge \ldots \wedge var_n \in range_n$ then
> Load the decision process generated in M8) for the S-Q level CART tree corresponding to L
> where $<var_0, range_0>, \ldots,$ and $<var_n, range_n>$ are all the 2-tuples in L.

The conditional evaluation statements generated in **S-7)** will then be embedded in our QoS monitoring modules. Similarly, the A-S level CART trees can also be simplified and merged. Hence, in runtime, the more detailed decision process generated in *S-6)* will only be loaded when a "QoS requirement not satisfied" event is generated and the corresponding condition for loading this decision process is met.

4.3. *Synthesizing WF Controllers*

The WF controllers in our approach mainly perform two tasks: (1) discovering replacement services when there are service, system or network failures, and (2) finding new service configurations to satisfy QoS requirements. Based on our coordination protocol presented in Section 4.1, the WF controllers to be synthesized should have the general form described in α-calculus as shown in Fig. 8.

In the α-calculus descriptions in Fig. 8, the strings started and ended with "$" are the placeholders for the names of WF controllers and communication channels among WF controllers, monitors and agents, and will be replaced by the actual names after all WF controllers, monitors and agents are synthesized.

A WF·controller takes the service name s, host name h, network name n, QoS requirements qr, and the initial system state *state* as the initial parameters. It first waits for messages from WF monitors and other controllers (line 2). *Failure* and *controlMessage* are the named channels for WF monitors and other controllers to send notifications of service/host/network failures, unexpected delays, and unsatisfied QoS requirements. *$STATUS$* is the named channels for the WF monitor responsible for s to send relevant system status information to the WF controller responsible for s. Lines 3–9 define how the WF controller handles service/host/network failures [see *Case C-1.1)* and *Case C-2.1)* in our coordination protocol presented in Section 4.1]. Lines 10–12 define how the WF controller handles unexpected delay caused by services invoked before s. In particular, in line 12, a reconfiguration handling sub-process is initiated with new QoS requirements, in which the maximum allowable delay for s is reduced. Lines 13–17 specify how the WF controllers handle unsatisfied QoS requirements of s, which are detected by the WF monitor for s.

In the WF controllers and their associated reconfiguration handlers (see Fig. 8), two system control services are used: *serviceDiscovery* and *findServiceConf.* The *serviceDiscovery* service takes four inputs: the failure detected (the name of the failed service/host/network or an empty string

```
1      fix $CONTROLLER$(string s, string h, string n, tuple qr, tuple state)=
2        ((string failure(f) + tuple controlMessage(string msg_id, tuple msg_val)) || tuple $STATUS$(tuple state)).
3        (if (f = s or f = h or f = n) then
4          let tuple r(string rs, string rh, string rn) = SystemControlService:serviceDiscovery(f, s, qr, state) instantiate
5            if (rs != "") then
6              tuple $SERVICE_REPLACEMENT$<s, rs, rh, rn>.
7              $CONTROLLER$ (rs, rh, rn, qr, state)
8            else
9              tuple controlMessage<"TERMINATION", null>
10         else if (msg_id = "UNEXPECTED_DELAY" and msg_val.source != s) then
11           (qr.max_delay << qr.max_delay - msg_val.delay).
12           $RECONFIGURATION_HANDLER$(s, h, n, qr, state)
13         else if (msg_id = "UNSATISFIED_QoS" and msg_val.source = s and msg_val.stage = "preExecution") then
14           $RECONFIGURATION_HANDLER$ (s, h, n, qr, state)
15         else if (msg_id = "UNSATISFIED_QoS" and msg_val.source = s and msg_val.stage = "inExecution") then
16           string $SUSPENDAGENT$("suspend").
17           $RECONFIGURATION_HANDLER$ (s, h, n, qr, state)
18         else if (msg_id = "TERMINATION") then
19           zero
20         else $CONTROLLER$(s, h, n, qr, state))

21     fix $RECONFIGURATION_HANDLER$(string s, string h, string n, tuple qr, tuple state) =
22       let string conf = SystemControlService:findServiceConf(s, h, n, qr, state) instantiate
23         if (conf != "") then
24           tuple $SERVICE_RECONFIGURATION$<s, conf>.
25           $CONTROLLER$(s, h, n, qr, state)
26         else let tuple r(string rs, string rh, string rn) = SystemControlService:serviceDiscovery("", s, qr, state) instantiate
27           if (rs != "") then
28             tuple $SERVICE_REPLACEMENT$<s, rs, rh, rn>.
29             $CONTROLLER$(rs, rh, rn, qr, state)
30           else
31             tuple controlMessage<"TERMINATION", null>
```

Fig. 8.   The general form of WF controllers.

if no failure has been detected), the service s to be replaced, the QoS
requirements qr that need to be satisfied by the newly discovered service,
and the current system status $state$. The $serviceDiscovery$ service will find
a new service rs to replace s, which has the same functionality as s and
can satisfy qr under $state$. The failure detected will be used to reduce the
search space since the $serviceDiscovery$ service will exclude the services
in the failed host or network. If rs is found, the $serviceDiscovery$ service
will return the name of rs, and the names of the host and network,
where rs is deployed. The $serviceDiscovery$ service can be implemented

based on existing standards for service discovery, such as UDDI (http://www.oasis-open.org/committees/uddi-spec/doc/tcspecs.htm) and WS-Discovery (http://docs.oasis-open.org/ws-dd/discovery/1.1/os/wsdd-discovery-1.1-spec-os.pdf), or other advanced service discovery mechanisms, such as our work presented in Ref. 32. Our current implementation is based on UDDI.

The *findServiceConf* service takes five inputs: the name of the service (s), the name of the host for $s(h)$, the name of the network used by $s(n)$, the QoS requirements qr that need to be satisfied by s, and the current system status *state*. In our current approach, we assume that a service s has finite number of configurations. The ASQ models for s are used in the *findServiceConf* service to estimate the possible QoS under different configurations. Based on the ASQ models, the *findServiceConf* service performs a search on all configurations of s to find a configuration *conf* that satisfies qr under *state*.

5. Conclusions and Future Work

In this chapter, we have presented a virtual machine-based architecture, called *WVM*, for workflow execution, monitoring and control in SBS, and a coordination protocol defining how distributed WF agents, monitors and controllers coordinate to execute, monitor and control the adaptation of workflows in SBS based on WVM. We have also presented our approach to synthesizing WF monitors and controllers for proactive monitoring and control of workflow execution in SBS. Our approach is expected to greatly reduce the effort of developing high-quality SBS, especially for large-scale SBS with many long-running services and workflows that need to satisfy multiple QoS requirements simultaneously. The WF monitors and controllers developed using our approach will provide proactive, autonomic, and efficient workflow monitoring and control in highly dynamic operation environments. Future work in this area includes the incorporation of task migration, development of new methods for finding service configurations to achieve better QoS, integrating our virtual machine-based architecture with our three-layered intelligent control architecture for ASBS[33] to provide better support for system adaptation, and performing simulations and experiments to examine the overhead and benefits of our approach and compare our approach with other existing approaches.

Acknowledgement

The work reported here was supported by the DoD/ONR under the Multidisciplinary Research Program of the University Research Initiative, Contract No. N00014-04-1-0723, and National Science Foundation under grant numbers CNS-0524736 and CCF-0725340.

References

1. Ponnekanti, S., Fox, A.: Sword: A developer toolkit for web service composition, 11th *Int'l World Wide Web Conf. (WWW 2002) Web Engineering Track*, 2002. Available at: http://www2002.org/CDROM/ alternate/786/.
2. Rao, J., Kungas, P., Matskin, M.: Application of linear logic to web service composition, *Proc. 1st Int'l Conf. on Web Services*, 2003, pp. 3–9.
3. Sirin, E., Hendler, J.A., Parsia, B.: Semi-automatic composition of web services using semantic descriptions, *Proc. Web Services: Modeling, Architecture and Infrastructure (WSMAI) Workshop* in conjunction with the 5th Int'l Conf. on Enterprise Information Systems (ICEIS 2003), 2003, pp. 17–24.
4. Nguyen, X.T., Kowalczyk, R., Phan, M.T.: Modeling and solving QoS composition problem using fuzzy DisCSP, *Proc. 2006 IEEE Int'l Conf. on Web Services (ICWS 2006)*, 2006, pp. 55–62.
5. Berbner, R., *et al.*, Heuristics for QoS-aware web service composition, *Proc. 2006 IEEE Int'l Conf. on Web Services (ICWS 2006)*, 2006, pp. 72–82.
6. Cai, K.-Y., Chen, T.Y. and Tse, T.H.: Towards research on software cybernetics, *Proc. 7th IEEE Int'l Symp. on High Assurance Systems Engineering (HASE'02)*, 2002, pp 240–241.
7. Cai, K.-Y., Cangussu, J.W., DeCarlo, R.A. and Mathur, A.P.: An overview of software cybernetics, *Proc. 11th Ann. Int'l Workshop on Software Technology and Engineering Practice*, 2004, pp. 77–86.
8. Yau, S.S., Ye, N., Sarjoughian, H., Huang, D.: Developing service-based software systems with QoS monitoring and adaptation, *Proc. 12th IEEE Int'l Workshop on Future Trends of Distributed Computing Systems (FTDCS 2008)*, October 2008, pp. 74–80.
9. Bacchus, F., Kabanza, F.: Using temporal logic to control search in a forward chaining planner, in *New Directions in Planning*, Ghallab M. and Milani A. (eds.), IOS Press, 1996, pp. 141–153.
10. Woodman, S.J., Palmer, D.J., Shrivastava, S.K., Wheater, S.M.: Notations for the specification and verification of composite web services, in *Proc. the 8th IEEE International Enterprise Distributed Object Computing Conference (EDOC'04)*, September 2004, pp. 35–46.
11. Jin, J., Nahrstedt, K.: On exploring performance optimization in web service composition, in *Proc. ACM/IFIP/USENIX Int'l Middleware Conf.*, October 2004, pp. 115–134.

12. Canfora, G., Di Penta, M., Esposito, R., Villani, M.L.: An approach for QoS-Aware service composition based on genetic algorithms, in *Proc. 2005 Conf. on Genetic and Evolutionary Computation*, 2005, pp. 1069–1075.

13. Guo, C., Cai, M., Chen, H.: QoS-Aware service composition based on tree-coded genetic algorithm, in *Proc. 31st IEEE Ann. Int'l Computer Software and Applications Conf. (COMPSAC 2007)*, July 2007, pp. 361–367.

14. Kandasamy, N., Abdelwahed, S., Hayes, J.P.: Self-optimization in computer systems via on-line control: Application to power management, in *Proc. 1st Int'l Conf. on Autonomic Computing*, May 2004, pp. 54–61.

15. Dong, X., Hariri, S., Xue, L., *et al.*, AUTONOMIA: An autonomic computing environment, in *Proc. IEEE Int'l Conf. on Performance, Computing, and Communications*, April 2003, pp. 61–68.

16. Tannenbaum, T., Wright, D., Miller, K., Livny, M.: Condor — A distributed job scheduler, in *Beowulf Cluster Computing with Linux*, T. Sterling (eds.), MIT press, 2002. Available at: http://www.cs.wisc.edu/condor/doc/beowulf-chapter-rev1.pdf.

17. Yu, J., Buyya, R.: A novel architecture for realizing Grid workflow using tuple spaces, *Proc. 5th IEEE/ACM Int'l Workshop on Grid Computing (GRID 2004)*, 2004, pp. 119–128.

18. Cao, J., Jarvis, S.A., Saini, S., Nudd, G.R.: GridFlow: Workflow management for Grid computing, *Proc. 3rd Int'l Symp. on Cluster Computing and the Grid (CCGrid 2003)*, 2003, pp. 198–205.

19. Ludäscher, B., *et al.*, "Scientific workflow management and the Kepler system", *Concurrency and Computation: Practice & Experience*, 18(10), 2006, pp. 1039–1065.

20. Deelman, E., *et al.*, Mapping abstract complex workflows onto Grid environments, *J. Grid Computing*, 1(1), 2003, pp. 25–39.

21. Cao, J., *et al.*, ARMS: An agent-based resource management system for Grid computing, *Scientific Programming*, 10(2), 2002, pp. 135–148.

22. Bharadwaj, R.: Secure middleware for situation-aware naval C^2 and combat systems, *Proc. 9th IEEE Int'l Workshop on Future Trends of Distributed Computing System (FTDCS'03)*, 2003, pp. 233–240.

23. Amir, Y., Nita-Rotaru, C., Stanton, J., Tsudik, G.: Secure spread: An integrated architecture for secure group communication, *IEEE Trans. on Dependable and Secure Computing*, 2(3), 2005, pp. 248–261.

24. May, D., Shepherd, R.: The transputer implementation of occam, *Proc. Int'l Conf. on Fifth Generation Computer Systems*, 1984.

25. Caromel, D., Henrio, L.: *A Theory of Distributed Objects*, Springer Verlag, 2005.

26. Milner, R.: *Communicating and Mobile Systems: The π-Calculus*, Cambridge University Press, 1999.

27. Yau, S.S., *et al.*, Automated agent synthesis for situation-aware service coordination in service-based systems, Technical Report, Arizona State University, 2005, available at: http://dpse.eas.asu.edu/as3/papers/ASU-CSE-TR-05-009.pdf.

28. Yau, S.S., *et al.*, Automated situation-aware service composition in service-oriented computing, *Int'l Jour. on Web Services Research (IJWSR)*, 4(4), 2007, pp. 59–82.
29. Cardelli, L., Gordon, A.D.: Mobile ambients, *Theoretical Computer Science*, 240(1), 2000, pp. 177–213.
30. Yau, S.S., Huang, D., Zhu, L.: An approach to adaptive distributed execution monitoring for workflows in service-based systems, *Proc. 4th IEEE Int'l Workshop on Software Cybernetics (IWSC)*, July 2007, pp. 211–216.
31. Hardwick, J., and Stout, Q.F.: Flexible algorithms for creating and analyzing adaptive sampling procedures, *New Developments and Applications in Experimental Design*, 34, 1998, pp. 91–105.
32. Yau, S.S., Liu, J.: Functionality-based service matchmaking for service-oriented architecture, *Proc. 8th Int'l Symp. on Autonomous Decentralized Systems (ISADS)*, 2007, pp. 147–154.
33. Chang-Hai Jiang, Hai Hu, Kai-Yuan Cai, Dazhi Huang, and Stephen S. Yau, An Intelligent Control Architecture for Adaptive Service-based Software Systems with Workflow Patterns, in *Proc. 32nd IEEE Ann. Int'l Computer Software and Applications Conf. (COMPSAC'2008)*, July 2008, pp. 824–829.
34. Yau, S.S., Ye, N., Sarjoughian, H., Huang, D., Roontiva, A., Baydogan, M., Muqsith, M.: Toward development of adaptive service-based software systems, *IEEE Transactions on Services Computing*, 2(3), 2009, pp. 247–260.

Chapter 9

ACCELERATED LIFE TESTS AND SOFTWARE AGING

RIVALINO MATIAS JR. ·

School of Computer Science
Federal University of Uberlândia, Brazil
rivalino@facom.ufu.br

KISHOR S. TRIVEDI

Department of Electrical and Computer Engineering
Duke University, USA
kst@ee.duke.edu

Accelerated life test (ALT) methods are successfully applied in many industries to reduce the test period of highly dependable products. Software industry is not different, having the same demand to reduce the period of test for software products with very low failure rates. Since software is now part of many important processes of modern life, increasingly software-based products are expected to be highly dependable, which requires sophisticated techniques to test them within acceptable time frames.

The use of ALT assumes that the system under test has its life reduced when it is exposed to some stress load. Many software products show a systematic performability and dependability degradation under certain circumstances, mainly when varying workload and long execution time are present. These software system degradations have been investigated over the last decade and now they are well explained through the software aging theory. In this chapter, we discuss how to apply accelerated life test techniques to software systems that suffer from software aging. We show how the software aging theory enables the usage of ALT methods to estimate the time to failure of a software system. We show a real case study, related to a widely adopted web server software system, to illustrate how to systematically apply ALT in software experimental studies. In this experimental study, ALT offers a reduction of approximately seven times the time required to obtain the same amount of failure data in use condition (without acceleration).

1. Introduction

Accelerated life test (ALT) methods are successfully applied in many industries to reduce the test period of highly dependable products. Software industry is not different, having the same demand to reduce the period of test for software products with very low failure rates. Since software is now part of many important processes of modern life, increasingly software-based products are becoming highly dependable, which require sophisticated techniques to test them within acceptable time frames.

The use of ALT implies that the system under test has its life reduced when it is exposed to some stress load. Its failure rate increases when the system is submitted to higher stress loading than it usually experiences in its use condition workload. Controlling the levels of stress loading reflects on the system's failure rate. ALT was originally created to be applied to products that have their lifetime governed by physical or chemistry laws. For these products, the stress is directly related to product's physical-chemical properties that degrade under certain circumstances, such as varying temperature, humidity, vibration, etc.

Similarly to the physical/chemistry-based products, many software systems show a systematic performability and dependability degradation under certain conditions, mainly when varying workload and long execution time are present. These software system degradations have been investigated over the last decade and now they are well explained through the software aging theory.[11] Due to the cumulative property of the software aging phenomenon, it occurs more intensively in continuously running processes[1] that are executed over a long period of time. Typically, aging-related failures are very difficult to observe because the accumulation of aging effects usually requires a long-term execution. Thus, collecting a significant sample of aging-related failure times in order to be used to estimate the system's lifetime distribution is a very hard task. This is an important problem that prevents many studies focused on analytical modeling of software aging aspects of using representative parameter values.

Due to the degenerative nature of the software aging phenomenon, in this chapter we discuss how to use accelerated life test techniques applied to software systems suffering from software aging. We present how the software aging theory enables the application of ALT methods to estimate the time to failure of a software system. This chapter is organized as follows.

Section 2 presents the most important aspects of the software aging theory, linking its main concepts with the fundamentals of software reliability engineering. Section 3 revisits the ALT methods, emphasizing how it can be applied to software systems with symptoms of software aging.

---

[1]An instance of a computer program being executed by the operating system kernel.

A detailed discussion is presented in this section in order to provide the necessary background for the rest of the chapter. Subsequently, in Section 4 we present a complete real case study, applying accelerated life tests to estimate the MTTF of a real software system during controlled tests. We evaluate the accuracy of the estimates against the observed failure times. We also present the reduction factor of the experimentation time when using accelerating life tests in comparison with controlled tests without life acceleration. Finally, a conclusion section will close the chapter bringing to the readers a few suggestions and insights about how to apply the above-mentioned discussions on industry and research projects.

2. Software Aging Theory

Since the notion of software aging was introduced fifteen years ago,[12] the interest in this phenomenon has been increasing from both academia and industry. The occurrence of software aging in real systems has been documented in the literature.[2,5,8] Many approaches have been used to study this phenomenon. The majority of these research efforts have concentrated on understanding its effects theoretically[3,23] and empirically.[15,24] Moreover, the search for mitigation resulted in the so-called software rejuvenation techniques.[12,27] This section discusses the foundations of the software aging phenomenon. We focus on conceptual and practical aspects involved, and present a set of definitions that we consider essential to understand the taxonomy for the software aging research.

Classical Software Failure Mechanics

In Ref. 1, the causal relationship between software's **fault, error** and **failure** is presented. A software system failure is observed when an internal error is propagated to the system interface and causes the service (output) delivered by the system to deviate from its specification. Typically, the system is considered failed only at the delivery time, that is, when its output (e.g., a numerical result) is received by the system's user and the unexpected (erroneous) output is detected. It is important to note that although the perception of error occurs at the output delivery time, the system may be in a failed state long before that. Thus, a failure is the perception of an error in the service provided by the system. An error is that part of the internal system state which may lead to a failure occurrence. If internal errors are not propagated to the system interface, the system is considered to be working correctly. Note that it is not necessary for multiple internal

errors to occur to have a system failure; even a single error occurrence
may be sufficient to lead to system failure. Errors can be transformed into
other errors. For example, an error in component 1 can reach the service
interface of this component, used by component 2, thus causing an error
in this second component. This transformation of errors is referred to as
error propagation. The error propagation leads to a system failure if
the error is propagated to the service interface of the system, causing the
service provided to deviate from its specification. Note that the term error
propagation is used for the transformation of an error into another error
both with and without the causation of a failure occurrence.

The cause of an error is the activation of a software fault. A software
fault is also known as a **defect** or a **bug**. There are several types of
faults, ranging from incorrectly coded instructions to high-level system
misconfiguration. A fault is considered dormant until its activation. For
example, an incorrect definition (e.g., *int* instead *long int*) of an accumulator
variable will lead to an internal error occurrence only when the variable
assumes *long int* values which overflows the storage capacity of an *int* type.
When such an event occurs, we have the fault activation. The activation of
a fault causes an internal error that may or not reach the system interface.
If it does, we have a **system failure**. The rate with which a dormant fault
will become active therefore depends heavily on the intensity and the way in
which a system is used; a quantitative characterization of the latter aspect
is the **operational profile**.[19]

Considering the **fault-error-failure chain** described above, we
conclude that the root cause of a software failure is a software fault.
However, if the fault is not activated, the chained effects that lead the
system to fail will not happen. Thus, the activation of a fault is the primary
event that leads to a software failure. Fault activation is the application
of an input (the **activation pattern**) to a code segment that causes a
dormant fault to become active. Most internal faults cycle between their
dormant and active states. Not all fault activations will cause a system
failure instantly thru the chain reaction explained above. The time between
the fault activation and the failure observation is mainly influenced by
the error propagation and the failure manifestation delays. The error
propagation delay is mainly caused by the computation process whereby
an error is successively transformed into other errors. As a result of the
computation process flow, the error state may move from its creation
point to the system interface in a slow manner. Related to the failure
manifestation delay, because of output buffers or similar mechanisms an

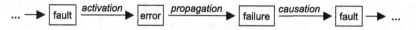

Fig. 1.   General fault-error-failure chain.[1]

erroneous result may not immediately be seen by system's users. Figure 1 presents the chain reaction cycles discussed above.

Avižienis *et al.*[1] present a scheme for classifying faults according to eight criteria, e.g., the system boundaries (internal or external), the phenomenological cause (natural or human-made), the objective (malicious or non-malicious), the persistence (permanent or transient), the dimension (hardware or software), and the phase of creation or occurrence (development or operation). Based on this classification scheme, faults in the software code, referred to as **software flaws** by Avižienis *et al.* and often simply called **software faults** or **(software) bugs**, can be described as internal human-made non-malicious permanent software development faults. Furthermore, Avižienis *et al.* mention a classification of faults according to their activation/propagation reproducibility: Faults whose activation and error propagation is reproducible are called **solid**, or hard, faults, whereas faults whose activation/propagation is not systematically reproducible are called **elusive**, or soft, faults. Especially for software bugs that permanently reside in the code (until they are detected and removed) it seems counter-intuitive that repeating the actions (like user inputs) that previously caused a failure will not lead to another failure when repeated. It is therefore of interest to study the properties that a software fault needs to feature in order to have the potential to be non-reproducible. However, the classification into solid and elusive faults is subjective, because it also depends on the knowledge of the respective user about the fault activation and error propagation mechanism of the fault in question, as well as on the operational behavior of the user. A specific fault could be considered "solid" by one user, but "elusive" by another one.

The definitions of Mandelbug and Bohrbug[9,10] classify the fault types using more objective criteria related to properties of the fault itself. A **Mandelbug** has the potential to be difficult to isolate and to cause failures that are not systematically reproducible. As an example, consider the code of an application in which the initialization of a variable is missing. If a debugger initializing all variables by default can prevent the fault from causing a failure, then this fault is Mandelbug, because the debugger, a part of the system-internal environment of the application, can affect fault activation. A **Bohrbug**, on the other hand, is an easily isolated fault that always manifests consistently under a well-defined set of conditions, because its activation and error propagation lack "complexity" as defined in Refs. 9 and 10. Bohrbug is the complementary antonym of Mandelbug. The Mandelbug definition uses the concept of the **(software-) system-internal environment** of an application. While the environment of an application consists of all the entities outside the system boundaries of the application (e.g., operating system, hardware, users, power supply network, office building), its system-internal environment only includes those entities outside the application that are located within the boundaries of the computer system. In particular, users and office infrastructure are excluded. The system-internal environment of an application thus contains the hardware, the OS, the other applications, etc.

Fundamentals of Software Aging

Aging-related (AR) failures are a specific type of software failure, so the concept of chain fault-error-failure discussed previously applies to it. Software aging is the name given to a phenomenon empirically observed in many software systems. It can be defined as a growing degradation of software's internal state during its operational life. A general characteristic of this phenomenon is the fact that, as the runtime period of the system or process increases, its failure rate also increases. Again, a failure can take the form of incorrect service (e.g., erroneous outcomes), no service (e.g., halt and/or crash of the system), or partial failure (e.g., gradual increase in response time). For physical systems, aging is well known to occur in the wear-out phase. In the classical bathtub curve[25] this behavior is illustrated by an increasing failure rate after a certain stable period of life. However, while hardware faults can come into existence due to wear and tear, it seems impossible, at first sight, that software bugs, which are permanent development faults, can be responsible for software aging.

Software aging is the practical consequence of software errors accumulation. What makes AR failure different from other software failure types is the concept of **error accumulation** in addition to error propagation. The notion of error accumulation is essential for characterizing the software aging phenomenon. Aging in a software system, similarly to hardware, or even in human beings, is an accumulative process. Accumulating the effects of successive error occurrences influences directly the AR failure manifestation time. Software aging effects are consequences of errors caused by aging-related fault activations. They work by gradually leading the

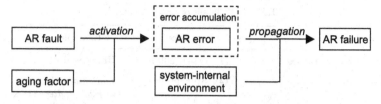

Fig. 2. Chain reaction for aging-related failures.[11]

system's erroneous state towards the failure occurrence. This gradual shifting as a consequence of aging effects accumulation is the fundamental nature of the software aging phenomenon. Figure 2 shows a modified version of the general *fault-error-failure* chain specific to the aging-related (AR) failures.

It is important to highlight that a system fails due to the consequences of aging effects accumulated over the time. For example, a database server system may fail due to insufficiency of available physical memory, which is caused by the accumulation of memory leaks caused by a software fault. In this case, the AR fault is a defect in the code (e.g., unbalanced use of malloc() and free() routines) that causes memory leaks; the memory leak is the observed effect of aging fault; the aging factors are those input patterns that exercise the code region where the AR fault is located, causing its activation. Based on the amount of leakage per AR error occurrence, and the availability of main memory (system-internal environment), a server system failure, due to the unavailability of memory, could manifest only after a long running time. Thus, the accumulation of AR error effects is the essential marker that indicates the presence of software aging in a given running system.

Different types of aging effects have been observed. Table 1 shows an initial set of aging effect classes based on their common characteristics.

Table 1 presents the four classes of aging effects that are the most cited in the literature. Resource leakage may occur in different ways. The important aspect about this class is the lack of de-allocation of previously allocated system resources, a well-known programming defect in large and complex system developing. Also, partial de-allocation contributes for leaking system resources. Memory leaking is the most recurrent and harmful type of aging effects observed so far. Important to note that, aging effects are present not only in user-level applications, but also inside the operating system kernel. Both software levels are prone to suffer from software aging. In terms of OS kernel, the device driver subsystem is the most vulnerable

Table 1.   Classes of aging effects.[11]

Basic class	Extension	Examples
Resource leakage	(1) OS-specific	— Unreleased
	(2) App-specific	• *Memory* (1, 2)
		• *File handlers* (1)
		• *Sockets* (1)
		— Unterminated
		• *Processes* (1)
		• *Threads* (1, 2)
Fragmentation	(1) OS-specific	— Phys. memory (1)
	(2) App-specific	— File system (1)
		— Database files (2)
Numerical error accrual	(1) OS-specific	— Round-off (1, 2)
	(2) App-specific	
Data corruption accrual	(1) OS-specific	— File system (1)
	(2) App-specific	— Database files (2)

to resource leak-based aging effects. This subsystem usually encompasses code produced by many third-party programmers, not being under the same quality control than other parts of OS subsystems.

In addition to the resource leak, fragmentation is the second largest cause of software aging. Fragmentation may also occur in both application and OS kernel levels. In application level, fragmentation, like resource leak, is restricted to the process's address space, affecting only the defective process. However, in kernel level, fragmentation will affect all running process under the OS. Fragmentation of database index files is a common example in application level. File system and main memory fragmentation are common cases in OS level, each one affecting the system's performance and availability in some extent. Section 4 presents a case study of aging caused by memory leaks in an application process (user-level). In terms of kernel-level aging research, Section 5 discusses some ongoing projects.

Aging effects can also be classified into volatile and non-volatile effects. They are considered volatile if they are removed by re-initialization of the system or process affected, for example via a system reboot. In contrast, non-volatile aging effects still exist after reinitializing of the system/process.

Physical memory fragmentation and OS resource leakage are examples for volatile aging effects. File system and database metadata fragmentation are examples for non-volatile aging effects. Another example of a non-volatile aging effect is numerical error accrual preserved between system reboots via checkpoint mechanism. Note that hibernation and similar mechanisms (e.g., standby), which preserve the system memory (and thus the aging effects present in it) between system reboot, allow the majority of intrinsically volatile aging effects to persist even after system/process re-initialization.

Aging effects in a system can be detected by monitoring aging indicators. Aging indicators are markers for aging detection, like antigens are markers to detect cancer disease. In the simplest approach, system health is considered a latent binary variable distinguishing between a stable internal state on the one hand and a failure probable state on the other.[11] Aging indicators are then explanatory variables that individually or in combination can suggest whether or not the system is healthy. They can be considered at several levels, such as OS, application process, application component, middle-ware, virtual machine (VM), and VM monitor (VMM). We can classify aging indicators in two general classes according to their granularity: **system-wide** or **application-specific**. System-wide indicators provide information related to subsystems shared by several running applications. Examples of shared subsystems are OS, middleware, VM and VMM, among others. Indicators in this category are often used to evaluate the aging effects on the system as a whole and not for a specific application, since the shared nature of their environment may cause noise in the captured data. Examples of aging indicators in this category are free physical memory, used swap space, file table size, and system load. Application-specific indicators provide specific information about an individual application process, thus giving more accurate information about it than system-wide indicators. When the application process is running under a VM (e.g., Java programs), then aging indicators applied to the VM can also be used as a reference for the application being executed under the VM. Examples of aging indicators in this category are resident set size of the process, Java VM heap size, and response time.

While in many cases software aging is due to AR bugs, even in the absence of such faults in the code aging effects can occur as a consequence of the natural dynamics of a system's behavior. This kind of aging is thus referred to as **natural aging**. Among the examples for natural aging are the fragmentation problems experienced by file systems, database index files, and main physical memory. For example, in database servers the fragmentation class of aging effects can be captured via aging indicators such as the degree of index-related metadata fragmentation (e.g., Tablespace fragmentation value in the Oracle DBMS). Considering not only the software system itself, but the higher-level system including its users, one could argue that natural aging is due to faults, namely to mistakes on the part of the operators; e.g., in the case of fragmentation problems the operator has made the mistake of not executing defragmentation routines. However, as such measures only mitigate the effects of natural aging, even "correct" behavior of the operators would not have solved the underlying problem. This is in contrast with software aging caused by AR bugs, discussed above, where fixing the software fault permanently removes the aging effect. The notion of natural aging without existence of a fault should not give the impression that any service degradation or any increase in the failure rate of a software system is considered software aging. Otherwise, this concept would also include increases in the failure rate that are merely due to changes in the operational profile or due to an increase in the intensity with which the system is being used.

In summary, we can conclude that the aging effect is not reversible without external intervention. For example, the accumulated internal error states caused by successive activations of aging-related faults do not disappear without external intervention; at the very best, no further errors may accumulate in the future, during periods in which the system is not exposed to any aging factors. Based on this characterization, an increasing failure rate due to the queuing of jobs in an overloaded system is not considered software aging, since the accumulated set of jobs not yet served will be reduced (and will finally disappear) once the workload falls below a certain threshold.

The time to aging-related failure (TTARF) is an important metric for reliability and availability studies of systems suffering from software aging. Based on the AR failure mechanisms explained above, we conclude that the probability distribution of TTARF is mainly influenced by the intensity with which the system gets exposed to aging factors; it is therefore influenced mainly by the system workload (and thus by the operational profile and the usage intensity of the system). Typically, aging-related failures are very difficult to observe because the aging effects accumulation process usually requires a long-term execution to expose the error to the system interface. Thus, collecting a significant sample of TTARF observations in order to be used to estimate the system's lifetime distribution is a very hard task. The rest of this chapter discusses how to use accelerated life tests to reduce the time to obtain the lifetime distribution of systems that fail due to software aging.

3. Accelerated Life Tests

Quantitative accelerated life tests (or simply ALT) are used in several engineering fields to significantly reduce the experimentation time.[21] ALT is designed to quantify the life characteristics (e.g., mean time to failure) of a system under test (SUT), by applying controlled stresses in order to reduce the SUT's lifetime and, consequently, the test period. Since the SUT is tested in an accelerated mode, and not in its normal operational condition, results have to be properly adjusted. ALT uses the lifetime data obtained under stress to estimate the lifetime distribution of the SUT for its normal use condition.

A fundamental element during the test planning is the definition of accelerating stress variable and its levels of utilization (load). Typical accelerating stresses are temperature, vibration, humidity, voltage, and thermal cycling.[21] These are appropriate for many engineering applications, where tests are applied to physical or chemical components that are governed by well-known physical laws. However, for software components we cannot adopt the above-mentioned accelerating stresses.

In order to apply ALT to software systems, we recall the discussion of Section 2. We assume that an AR failure will be observed as soon as the aging effects sufficiently accumulate for the delivered service to deviate from its specification. We also know that an error occurrence that produces the aging effects is caused by the activation of AR faults. Thus, we see that accelerating the activation of AR faults will reduce the time to AR failure being observed at the system interface. In order to accelerate the AR fault activations, we first need to identify the AR fault activation patterns and then increase their intensities. We assume that once we know the AR fault activation pattern(s) for the SUT, we can manipulate the SUT's workload to increase the rate of AR fault activations. Given the nature of AR faults, we need to determine suitable accelerating stresses based on pilot experiments. To date, software reliability engineering literature does not have standards, related to software accelerating stresses for ALT, such as those that exist in other engineering fields, so it is recommended an experimental approach to determine these stress factors.

In addition to the identification of accelerating stresses, a stress-loading scheme is also required. A stress-loading scheme defines how to apply the stress to the system under test. Two possible stress-loading schemes are: constant stress (time-independent), and varying stress (time-dependent).[16,21] According to Ref. 21, theories for the effect of time-varying stress on product life are in development, and mostly unverified. Thus, employing constant stress is recommended.

Once the SUT is tested at the selected stress levels, the observed failure times are for SUT operating under stress and not under its normal use condition — this is what reduces significantly the test time. Hence, the experimenter needs a model that relates the failure times observed at the tested stress levels to the underlying lifetime distribution of the SUT operating in its normal use condition. This model is called life-stress relationship.[16,17,21] Figure 3 illustrates this relationship. It shows two lifetime densities (shaded) built using failure times obtained with SUT under two higher stress levels as well as the estimated density for the SUT in its normal use condition. The normal-use-condition lifetime density (striped) is estimated through the life-stress relationship model, which relates a life characteristic of the SUT (e.g., mean life or MTTF) in high stress levels to its normal use condition level.

Several life-stress relationship models have been developed for different engineering fields. Examples of such well-known models are Arrhenius, Eyiring, Coffin-Manson, Peck, and Zhurkov.[16,17,21] Based on the SUT's physical/chemical properties, the underlying theories used to build these models assume specific stress types, for example, temperature for Arrhenius,

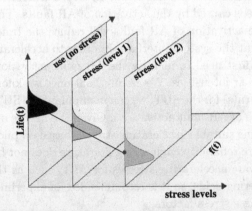

Fig. 3.   Estimated lifetime densities through a life-stress accelerated model.

humidity for Peck, and thermal cycles for Coffin-Manson. For this reason, traditional models applied to physical systems cannot justifiably be employed to build life-stress relationship models for software components. An exception is the Inverse Power Law (IPL) relationship model,[21] which has been successfully used in several ALT studies. The IPL is applicable to any type of positive stress, unlike the above-mentioned models that are used for specific types of stress variables. IPL is generally considered to be an empirical model for the relationship between life and the level of certain accelerating stress variable, especially those that are pressure-like stresses.[17] Equation (1) below is the IPL model.

$$L(s) = \frac{1}{k \cdot s^w},$$ (1)

where L represents a SUT life characteristic (e.g., mean time to failure), s is the stress level, k ($k > 0$) and w are model parameters to be determined from the observed sample of failure times.

The main property of this model is the scale invariance, that is, scaling s by a constant k simply causes the proportionate scaling of the function $L(s)$. It leads to a linear relationship when life and stress are plotted on a log-log scale:

$$\ln(L) = -\ln(k) - w \ln(s).$$ (2)

Figure 4 illustrates the fitting of the IPL model to sample of failure times, where 4(a) shows the original lifetime dataset for each stress level, and 4(b) the log-log plot of the dataset with a fitted linearized IPL model.

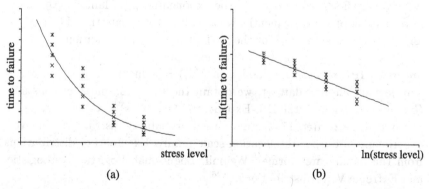

(a)   (b)

Fig. 4. (a) Life-stress relationship curve; (b) Linearized IPL model fitted to the lifetime logarithmized dataset.

Once (2) is fitted to the logarithmized lifetime dataset, we are able to estimate the values for parameters k and w, which are the straight-line's y-intercept and slope, respectively. These values are obtained in the logarithm scale, so we need to transform them before handling them off to (1). Given k and w in the correct scale, we use (1) to estimate the mean time to failure of the SUT for any value of s. Assuming that the s value equals to the SUT's use condition, (1) yields a point estimate of the MTTF for the SUT under normal operating regime.

In addition to the point estimate given by the life-stress model, experimenters are interested in confidence intervals[26] for the mean life of the SUT under normal use condition. Therefore, we need to combine the life-stress relationship with a probabilistic component in order to capture the variability of the dataset. This component follows a probability density function (pdf) and is independent of the stress variable. Once the pdf for the time to failure dataset is obtained, its parameters are estimated for each stress level.

The integration of the chosen life-stress relationship (e.g., IPL) with the fitted density function is necessary to relate them across the stress levels. Figure 3 illustrates this relationship. Essentially, after selecting the lifetime distribution and the stress-life relationship model, we make the distribution function parameter that characterizes the SUT's mean life (or also called characteristic life) dependent on the stress variable. For example, assuming that the SUT follows an exponential failure law, and the life-stress relationship used is IPL, we have an IPL-Exponential relationship model. The combination of both models is possible making SUT's mean time to failure, in this case $1/\lambda$, equal to $L(s)$ (see Eq. 1). As a result, we have $\lambda = ks^w$, which allows us to estimate the exponential pdf's lambda parameter as a function of s. Consequently, we are able to estimate the MTTF $(1/\hat{\lambda})$ and its confidence interval for the SUT at normal use condition.

Based on this parameter estimation approach, that is, by fitting the linearized IPL relationship model, $\ln(\lambda(s)) = -\ln(k) - w\ln(s)$, to a log-transformed lifetime dataset, we obtain the point estimates of constants (k and w) to use into the IPL-Exponential pdf, $f(t, s) = ks^w e^{-ks^w t}$.

According to Ref. 17, the most used probability distributions in ALT experiments are from the location-scale family. Examples of distributions from this family are Normal, Weibull, Lognormal, Logistic, LogLogistic, and Extreme Value distributions.[17,26]

Location-scale distributions have an important property in analyzing data from accelerated life tests, which is related to their cumulative distribution function (*cdf*). A random variable, Y, belongs to a location-scale family of distributions if its *cdf* can be written as:

$$Pr(Y \leq y) = F(y; \mu, \sigma) = \Phi\left(\frac{y - \mu}{\sigma}\right), \tag{3}$$

where $-\infty < \mu < \infty$ is a location parameter, $\sigma > 0$ is a scale parameter, and Φ does not depend on any unknown parameters. Appropriate substitution in Ref. 17 shows that Φ is the *cdf* of $(Y - \mu)/\sigma$ when $\mu = 0$ and $\sigma = 1$. The importance of this family of distributions for ALT is due to the assumption that the location parameter, in (3), depends on the stress variable, s, that is $\mu(s)$, and the scale parameter, σ, is independent of s. This relationship is shown in (4),

$$Y = \log(T) = \mu(s) + \sigma\varepsilon, \tag{4}$$

where ε is a probabilistic component modeling the time to failure sample variability. Essentially, we have a location-scale regression model to describe the effect that the explanatory variable, s, has on the time (to failure). Examples of ALT models for more than one explanatory variables and non-constant scale parameter may be found in Refs. 17 and 21. In the case study (Section 4) we evaluate density functions from location-scale family of distributions, and assume their scale parameter approximately constant (within the same confidence interval) across the stress levels.

Once the experimenter has identified the appropriate accelerating stress, the stress-loading strategy, and an approach to obtain the *pdf* combined with a stress-life relationship model, it is necessary to plan the accelerated tests in order to measure the failure times in the selected stress levels. The ALT experimental planning requires the definition of the following aspects: number of stress levels, levels of the stress variable, sample size, and the allocation proportions.

The number of stress levels is defined based on the ALT goals and restrictions. At least, two levels are necessary for ALT.[16] There is no theoretical limit on the number of stress levels but could be restricted by the experiment constraints (e.g., time, cost, etc.). The ALT literature (e.g., Refs. 16, 17 and 21) usually mentions three or four levels.

Clearly, the levels of stress variables should not be outside the SUT design limits. Using levels of stress surpassing these limits may introduce failure modes that are not present in SUT's normal operation. The main idea is to have the SUT operating under stress, thus outside the SUT's normal operating specification, but inside its design limits.

The sample size is related to the total number of tests, n, over all the stress levels. It is calculated based on the desired confidence intervals for the estimates.

The specification of the number of stress levels, the amount of stress applied at each level, the allocation proportion in each level, and the sample size follows one of the three most commonly used test plans[17]: traditional, optimal, and compromise plans. The traditional plans usually consist of three or four levels, equispaced, with the same number of replications (test units) allocated per level. The optimal plans specify only two levels of stress: high (S_H), and low (S_L). Meeker and Escobar[17] stated that the S_H value should be the maximal allowed within the design limits; and the S_L value, and its allocation proportion, π_L, should be selected to minimize the variance of the estimators of interest. Nelson[21] suggests that this allocation be based on the fraction p that minimizes the variance of the estimator at the use level of the stress variable. Assuming a sample size n, the number of allocated replications in S_L is the closest integer value of $(n \times p)$, where the remaining tests are then allocated to S_H. The compromise plans usually work with three or four stress levels, non-equispaced, and use an unequal allocation proportion. An example of a well-known compromise plan is the Meeker–Hahn plan,[21] which considers three stress levels, and follows an allocation proportion rule of 4:2:1. This allocation specifies, for a sample of n units, $4n/7$, $2n/7$, and $n/7$ test units allocated, respectively, to S_L, S_I, and S_H, where S_I is an intermediate stress level. In general, the compromise plans adopt a value of S_H based on practical aspects of the SUT, mainly the design limits. The S_I is equal to $(S_L + S_H)/2$, assuring that the levels are equidistant. Thus, the S_L value has to be specified to calculate the S_I, which according to Ref. 21 should be chosen taking into account the required accuracy for the estimates studied, at use level. A more detailed description of the three plans can be found in Refs. 17 and 21. In the case study of Section 4 we use a traditional plan, because in addition to decrease the SUT's lifetime we also want to know if the aging behavior is the same for different stress levels, thus using the same number of tests per level is necessary.

Accelerated Degradation Tests

Accelerated degradation tests (ADT) are a specific class of accelerated tests and extend ALT incorporating concepts from degradation tests.[17] The design of high-reliability systems implies failures are rare events, even during long periods of execution under high workload. For such highly reliable systems, an alternative approach to ALT are accelerated degradation tests, which do not look for failure times, but instead for a degradation measure of a product's performance characteristic taken over time, and under specific stress conditions.[22] In software aging research, the current difficulty to experimentally or empirically observe times to failure caused by software aging is similar to the highly reliable systems case mentioned above. Because ADT techniques were designed for physical systems, like in ALT, it is necessary to establish a mapping between core concepts in software aging theory and ADT methods to make it applicable to software aging studies. Basically, this mapping considers the same elements discussed previously for ALT, however including the measurement of the degradation caused by the software aging effects as part of the ADT model.

ADT has some advantages over ALT because performance degradation data can be analyzed sooner, even before any experimental units fail.[21] Also, performance degradation can yield better insights into the degradation process. However, such advantages can be achieved only if one has a suitable degradation model that establishes the relationship between the system degradation, and the accelerated stress variables.[17] According to Ref. 21, four common assumptions are adopted by the current degradation models: (a) degradation is not reversible, and performance always gets monotonically worse; (b) usually, the model applies to a single degradation process; and in case of simultaneous degradation processes, each one requires its own model; (c) degradation of a unit's performance before the test starts is insignificant; and (d) performance is measured with negligible random error.

In Ref. 17, the degradation path of a particular unit over time is denoted by $D(t)$, $t > 0$. A random sample of n test units are observed at pre-specified times $t_1, t_2, \ldots, t_s$. For each inspection, a performance measurement is registered for each test unit, and referred to as y. The inspection times are not required to be the same for all units, or even equidistant. Consider t_{ij} as the jth time measurement or inspection of the ith unit. The observed degradation measurement in unit i at time t_{ij} is represented by y_{ij}; and at the end of the test, the degradation path is registered as pairs $(t_{i1}, y_{i1}), (t_{i2}, y_{i2}), \ldots, (t_{im_i}, y_{im_i})$, for $i = 1, 2, \ldots, n$. The observed sample degradation y_{ij} of unit i at time t_{ij} is the unit's actual degradation, plus measurement error, and is given by

$$y_{ij} = D_{ij} + \varepsilon_{ij}, \quad i = 1, 2, \ldots, n \quad \text{and} \quad j = 1, \ldots, m_i \qquad (5)$$

where $\beta = (\beta_{1i}, \ldots, \beta_{qi})$ is a vector of model parameters which has dimension q, and $\varepsilon_{ij} \approx N(0, \sigma_\varepsilon)$ is a residual deviation of the ith unit at time t_j. The deterministic form of $D(t)$ is usually based on empirical analysis of the degradation process under study. The vector β corresponds to q unknown effects which determine the degradation path of unit i in the t_{ij} measures. Typically, a path model will have $q = 1, 2, 3$, or 4 parameters.[17] Some of these parameters can vary from unit to unit, and others could be modeled as constant across all units.[22] It is reasonable to assume that the random effects of the vector β are s-independent of the ε_{ij} deviations. Also, it is assumed that the ε_{ij} deviations are i.i.d. for $i = 1, \ldots, n$, and $j = 1, \ldots, m_i$. Due to the fact that the y_{ij} are measured sequentially in time, there is, however, a potential for autocorrelation among the ε_{ij}, $j = 1, \ldots, m_i$ values, especially when there are many closely spaced observations. As stated in Ref. 17, in many practical situations involving inference on the degradation of units, if the model fit is good, and if the testing and measurement processes are in control, the autocorrelation is typically weak. It is dominated by the unit-to-unit variability in the β values, and thus autocorrelation can be ignored. In the general degradation path model, the proportion of failures at time t is equivalent to the proportion of degradation paths that exceed the critical level D_f at time t. Thus, it is possible to define the distribution $F(t)$ of time-to-failure T from (5) as

$$F(t) = Pr\{T \le t\} = Pr\{D(t, \beta_1, \ldots, \beta_q) \ge D_f\}. \qquad (6)$$

For fixed D_f, the distribution of T depends on the distribution of $\beta_1, \ldots, \beta_q$. $F(t)$ can be expressed in closed form for simple degradation models. For complex models, specially when $D(t)$ is nonlinear, and more than one of the parameters is random, it is necessary to evaluate $F(t)$ with numerical methods.

In Ref. 3, the software aging phenomenon is defined as the continuous, increasing deterioration of the process' internal state, or as the degradation of system resources. In the former case, because the phenomenon is confined to the process memory (internal state), measuring its progress is difficult. For example, consider the accumulation of round-off errors in a global numeric variable inside a process memory image. In this case, the monitoring of aging should be possible if the program code is instrumented. When this cannot be easily implemented (e.g., software system is composed of closed third-party software components), the monitoring of aging alternatively could be possible through the individual process' performance measures, and resource consumption observable externally. For the later case of degradation system resources, one can follow the software aging evolution through the monitoring of operating system resources. Usually, ADT can be applied to any scenario where significant measures of the process/system degradation are observable. The experimenter can consider the degradation path $D(t)$ the progress of the aging effects on the system under test (SUT).

As discussed before, the main differences between applying ALT/ADT techniques to software aging experiments in comparison with other research fields (e.g., materials science) are the definition of degradation mechanisms, and how to accelerate such mechanisms. In other areas, the degradation mechanisms (e.g., wear) are usually related to physical-chemical properties of the SUT, which are used for failure/degradation acceleration purpose. In the case of software aging, the degradation mechanisms can be understood as the degenerative effects caused by the activation of software faults related to software aging. Hence, in systems that display software aging effects, it is possible to accelerate the degradation by use-rate, and by overstressing. The first can be achieved by increasing the frequency of the system usage to reduce the aging factors (A_F) latency. A similar technique was adopted in Refs. 4 and 7. In some cases, increasing the usage rate is not sufficient due to the low probability of the occurrence that the A_F has in relation to the remaining operations from the SUT operational profile.[20] Hence, in addition to increasing the A_F use-rate with respect to the SUT use condition, another option is to consider the accelerated degradation by overstressing. The A_F should be defined to allow different levels of influence in the SUT degradation acceleration. If it is necessary, the A_F could also be combined with other secondary factors (e.g., environmental factors[11]) to provide multiples stress levels. Individually or combined, the control of the frequency (use-rate), and intensity (stress level) of the A_F is achieved through the system workload. Therefore, we consider the A_F as one of the synthetic workload parameters used during the ADT.

Unlike the other areas that usually are based on physical laws for the accelerating stress definition, the selection of the A_F should be based on a sensitivity analysis of the aging phenomenon with respect to workload parameters at use condition. Given the chosen method of workload characterization, a practical approach is to use the statistical design of experiment (DOE)[13,18] for the A_F selection. In this case, the experimenter should consider as the DOE response variable (y) a measure that indicates the level of system aging during each run execution.[13] In Ref. 18, Montgomery suggests several experimentation strategies such as best-guess, one-factor-at-a-time, and factorial. The correct approach in dealing with several factors is to conduct a factorial design because all factors are varied together instead of one at a time, this being an efficient method to study the effects of two or more factors on the response variable. This strategy is thus chosen for the A_F identification. Among its variants (two-factor, fractional, 2^k, mixed-levels, etc.), the 2^k factorial design is particularly useful in factor screening experiments,[18] and thus is suited for studies focused on the aging characterization to identify the A_F. Thus, the A_F could be considered the workload parameter that contributes the most to the increase in the SUT aging effects. When more than one parameter, individually or through interactions, have significant influence over aging, the A_F will be their combination. In a combined form, the A_F can be seen as the operational mode[20] that causes greater influence on the system's aging effects.

Using this approach, the experimenter is able to maximize the aging acceleration through the A_F control inside the workload. This control is important because, in many practical situations, a high workload does not guarantee the aging acceleration, because the operational mode used could not create the necessary conditions for the activation of aging-related faults. For this reason, the workload characterization, and the sensitivity analysis of its parameters on the aging effects, are fundamental for the correct A_F selection. In Ref. 13, several techniques are presented to support the workload characterization, such as principal component analysis, multi-parameter histograms, clustering analysis, and Markov models. Each one of these techniques deals with specific requirements, and based on the experimenter's objectives they are appropriately selected.

In addition to the aforementioned ADT elements, another important quantity to be specified is the D_f threshold (see Eq. 6). This value depends on the specific characteristics of the SUT, as well as on the experimenter's goals. The instrumentation adopted to measure the degradation evolution until D_f usually depends on the type of aging effects is being monitored. For example, if the aging effects are application-specific (see Section 2), so a user-level monitor is sufficient. However, if it is system-wide, then a kernel-level monitor may be necessary. In any case, the inspection of y_{ij} is taken in several repetitions during the ADT, and not just at the end of the test as in ALT. The number of repetitions (m_i) of these measurements should be separated in time to minimize possible autocorrelations among ε_{ij} for

$j = 1, \ldots, m_i$ values.[17] If the measured degradation crosses the D_f threshold even though the SUT still up and running, then it is declared failed and the failure time is called pseudo-failure time. The SUT also can fail before its degradation level reaches the D_f. In both cases, the sample of pseudo-failure times and failure times are analyzed as a single sample of failure times.

As in ALT, ADT requires a model that relates the system's degradation observed in the evaluated stress levels to estimate a proper underlying lifetime distribution for the SUT's use condition. This model is called the stress-accelerated degradation relationship,[21] which in the context of software aging it translates into the term stress-accelerated aging (SAA). Such a model is usually based on traditional ALT models, such as Arrhenius, Eyiring, Coffin Manson, Inverse Power, etc. As in ALT, due to the lack of equivalent models established for ADT applied to software experiments, a natural candidate is the Inverse Power Law (IPL) as discussed earlier in this section.

Once the SAA is established, the next step is to estimate the underlying lifetime distribution for each stress level, and then to use them to estimate the $F(t)$ for the use condition. First, the experimenter needs a sample of failure times or pseudo-failure times for each stress level. For those degradation paths whose failure times are observed $(D(t) \geq D_f)$ within the test period, the failure times sample is taken directly. Otherwise, it is used the accelerated degradation data set from each degradation path to establish an accelerated degradation model, $D(t)$, and then to estimate pseudo-failure times. The $D(t)$ model can be (but need not be) the same for each degradation path. In ADT, steps to estimate the lifetime distribution, called the approximation method,[17,22] are as follows.

i. For the chosen stress levels, fit the model $y = D(t) + \varepsilon$ for each unit i. The model effects are considered as fixed for each unit, and random across them.

ii. Estimate the vector $\hat{\beta}_i = (\hat{\beta}_{i1}, \ldots, \hat{\beta}_{iq})$ for the unit i by means of the least-squares method.

iii. Solve the equation $D(t, \hat{\beta}_i) = D_f$ for t, and call the solution $\hat{t}_i$.

iv. Repeat the procedure for each sample's path to obtain the pseudo-failure times $\hat{t}_1, \ldots, \hat{t}_n$ for that stress level.

v. To the samples of failure or pseudo-failure times, apply the usual lifetime data analysis[17] to determine the $\hat{F}(t)$ for each stress level.

Through the SAA relationship previously established, and the $F(t)$ estimated for each stress level, the experimenter obtains the $\hat{F}(t)$ for the use condition. Several dependability metrics can be estimated for the system under test, once we have estimated the $F(t)$ for its use condition level. In addition, based on the $\hat{F}(t)$ for the use condition, software rejuvenation[27] mechanisms can be applied proactively to prevent system failure due to software aging effects. For example Ref. 6 discusses algorithms to obtain the optimal software rejuvenation schedule based on the closed form for $F(t)$.

4. Case Study

In this section we present a practical use of ALT in an experimental study of software aging. The system under test is the Apache web server, henceforth called httpd, a largely adopted web server system.[28] Software aging effects on the httpd have been previously investigated and reported in the literature.

In Ref. 8, the aging effects in httpd were measured and modeled as the degradation of SUT's main memory. By contrast, Ref. 15 measured the increase in httpd process size (httpd's resident set size — RSS) and verified that it was a consequence of memory leaks in the httpd processes. The effects of successive memory leaks in application processes can cause problems ranging from unaccepted response time (due to virtual memory thrashing) to system hang/crash caused by memory exhaustion. Particularly on httpd, both the above problems were experimentally observed in Ref. 15. On httpd-based web server systems, the accumulated effects of memory leaks are especially intensified because the same faulty program (httpd) is executed by many processes at the same time. It is not uncommon for such systems to run more than 300 httpd processes simultaneously. Figure 5 shows a memory snapshot (top command output), taken from our test bed, during a preliminary accelerated test with httpd.

```
PID   USER    PRI  NI SIZE   RSS  SHARE STAT %CPU %MEM  TIME COMMAND
21566 daemon   9   0 40368  40M   2684  S     0.0  3.2  0:25 httpd
21563 daemon   9   0 30924  26M   5144  S     0.0  2.6  0:04 httpd
21826 daemon   9   0 30616  25M   4180  S     2.1  2.5  0:04 httpd
21565 daemon   9   0 31640  24M   2092  S     0.3  2.4  0:04 httpd
21564 daemon   9   0 30728  22M    844  S     0.0  2.2  0:07 httpd
21567 daemon   9   0 12236  10M   2132  S     0.0  1.0  0:05 httpd
21917 daemon   9   0  6256 7088   2668  S     0.0  0.6  0:00 httpd
21713 daemon   9   0  6200 7034   2668  S     0.0  0.6  0:00 httpd
```

Fig. 5. Aging effects on httpd processes.

Figure 5 lists just a few httpd processes, sorted by their resident set sizes (RSS) (in descending order). Initially, all httpd processes started with approximately 6000 KB. Process 21566 showed a significant increase in its memory size (RSS) in comparison with the other httpd processes. In this test, we controlled the exposition of httpd processes to the factor that causes memory leaks, and intentionally exposed process (21566) more than others. If we assume that in a web server system all 300 httpd processes are exposed to the same aging factor and each httpd reaches 40 megabytes of RSS, then it is necessary to have at least 11 gigabytes of RAM to keep them all running. In fact, much more memory is required considering OS kernel and other administrative processes loaded in the same system, and that swapping httpd's pages is strongly discouraged due to its high negative impact on the httpd's response time.[15]

The rest of this section applies the procedures discussed in Section 3 to set up the experimental plan. First, we define the stress variable. We use the results presented in Ref. 15, which experimentally demonstrated that the HTTP request type, specially requests addressing dynamic content, and the page size were the most important factors in causing memory leaks in the SUT. Hence, the SUT's AR fault activation pattern is considered to be the arrival rate of requests addressing dynamic content. Increasing or decreasing the size of dynamically generated pages, in the SUT's workload, is how we control the stress loading on the SUT. Different from many physical systems under accelerated tests, which are influenced not only by the stress variable but also by the test time (e.g., lithium battery), in software components suffering from aging the runtime is not relevant since their failure mechanisms are not governed by physical/chemical laws. For example, in our preliminary acceleration experiment (see Figure 5), one could erroneously infer that process 21566 is bigger because of its higher CPU time (column TIME). CPU time or even runtime (process' uptime) is not necessarily correlated to the amount of process aging. If a process runs for a long period of time, but it is not exposed to aging factors (AR fault activation patterns), then it will not manifest any sign of aging effects. For example, process 21567 has higher CPU time than processes 21563, 21826 and 21564, however it has half the resident set size. As discussed in Section 2, what makes aging effects accumulate in a process is the degree of exposition of that process to aging factors.

In terms of stress loading strategy, based on the explained in Section 3, we use constant stress. We control it by handling the SUT's workload. It is possible to set the workload generator tool to issue HTTP requests addressing specific sizes of dynamic pages. This control is done in each stress level. For the life-stress relationship, we use the IPL model. We choose IPL based on its flexibility to model any type of pressure-like positive stress, that is compatible with our workload-based stress loading strategy and stress variable.

We employ three stress levels derived from the average page size of the SUT in its normal operating regime (use rate). In Ref. 15, this page size value is approximately 200 kB. Therefore, we adopted two, three, and four times this reference value as the ALT stress levels. For short, we call them $S1$ (400 kB), $S2$ (600 kB), and $S3$ (800 kB), respectively.

This experimental plan follows a traditional arrangement, so each stress level is tested equally resulting in an allocation proportion of $n/3$. The number of tests, n, is not defined in terms of test units, such as in accelerated tests applied to physical systems, but in number of replications. For software components, there is no difference among test units, given that the code tested is the same (httpd binary file). The difference among test replications is the system environment. Although each replication executes the same code (the httpd program), the environmental aspects (e.g., operating system, networking, and hardware) influence the test results for each replication. Even though the workload is accurately controlled, the randomness coming from the system environment has to be considered to estimate the mean life for the SUT at use rate. For this reason, it is more appropriate to refer to the number of test replications than the sample size. The number of replications is calculated using the algorithm proposed in Ref. 21. This algorithm is applicable when the accelerated lifetime dataset is used to estimate the mean and/or median of the failure times. For different estimates appropriate algorithm modifications are discussed in Ref. 21. The steps of the algorithm are described below.

First, a pilot sample of at least 21 failure times is required. Our experimental plan is based on three stress levels, thus it requires 7 failure times per level. After obtaining the pilot sample, we determine the best-fit *pdf* for the sample. Assuming that the fitted random variable will be one of the three most common density functions used in ALT (lognormal, Weibull, or exponential), the dataset must be transformed with $\log_{10}$ (for Weibull and lognormal), or $\log_e$ (for exponential). Subsequently, the number of replications (sample size) is calculated solving Eqs. (7) to (11).

$$\bar{x} = (n_1 x_1 + \cdots + n_j x_j)/np, \qquad (7)$$

where x_j is the value of the stress level transformed ($\log_{10}$ or $\log_e$) according to the above cited rule; n_j is the number of tests (replications) executed in the jth level of stress, and np is the total number of tests conducted for the pilot sample $(np = n_1 + \cdots + n_j)$.

$$\bar{y}_j = (y_{1_j} + y_{2_j} + \cdots + y_{n_j j})/n_j, \tag{8}$$

where $y_{n_j j}$ is the transformed ($\log_{10}$ or $\log_e$) value of jth failure time obtained in the level of stress j.

$$s_j = \left\{ \frac{[(y_{1j} - \bar{y}_j)^2 + \cdots + (y_{n_j j} - \bar{y}_j)^2]}{v_j} \right\}^{1/2}, \tag{9}$$

where v_j ($v_j = n_j - 1$) is the degrees of freedom of s_j, and s_j is the standard deviation of failure times obtained at jth stress level.

$$s = \left[\frac{(v_1 s_1^2 + \cdots + v_j s_j^2)}{v} \right]^{1/2}, \tag{10}$$

where v is the number of degrees of freedom calculated as $v = v_1 + \cdots + v_j$, and s is the pooled estimate of the log-transformed ($\log_{10}$ or $\log_e$) standard deviation (σ) in (11).

$$n_{ALT} = \left\{ 1 + (x_0 - \bar{x})^2 \left[\frac{np}{\sum (x - \bar{x})^2} \right] \right\} \left(\frac{z_{\alpha/2}\sigma}{\zeta} \right)^2, \tag{11}$$

where x_0 is the log transformation ($\log_{10}$ or $\log_e$) of the stress value assumed at normal use rate, and z is the tabulated value for the standard normal distribution at a given significance level (α); ζ is the precision of the estimate, which depends on the fitted *pdf* and the metric of interest for the accelerated failure times, where $\zeta = r$ (for mean) or $\zeta = \{ \log_{10}(r)$ or $\log_e(r) \}$ (for median); r is the precision for the estimator of interest. When $\zeta = r$, r is the half-width of the interval used to calculate the confidence interval for the mean. Alternatively, $r = (1 + m)$, where $m \times 100\%$ is the tolerated error for the estimator of the median. Finally, n_{ALT} can be computed and if its value is greater than the size of the pilot sample, then it is necessary to run more $(n_{ALT} - np)$ replications, rounded up to be able to equally distribute the additional tests among the three stress levels.

Numerical Results

Before analyzing the numerical results, it is important to define the notion of failure adopted in this study. In Ref. 15, tests containing dynamic requests lead the amount of memory used by httpd processes to reach the limit of the total main memory available. It forced the Linux kernel (virtual memory subsystem) to aggressively do swapping for more than thirty minutes. This unstable condition caused a dramatic increase of SUT's response time with many timeout failures observed on the client side. The accuracy of the monitoring instrumentation is negatively impacted by high swapping rate, which reduces the sample quality. To avoid the above-mentioned problems, we test an individual httpd process. It is similar to the test showed in Figure 5, where process 21566 was exposed to aging effects to a larger extent. In this case, the target httpd process is under different levels of stress, which means receiving requests for dynamic content at different arrival rates. The SUT is considered failed when the size of the httpd process crosses 100 megabytes. The rationale is that our SUT has 3 GB of RAM, so assuming a typical deployment of 200 httpd processes,[15] where about 15% of these processes are equally exposed to aging effects (e.g., requests for dynamic content of similar size), a total of 30 aged processes with 100 MB each is sufficient to cause a system failure due to memory saturation.

In addition to the failure definition, we also introduce our notion of time to failure. We measure the size of the httpd process after every 100 requests. Hence, we consider the time to failure not the wall-clock time, but the number of bunches of 100 requests processed before the httpd size crosses the specified threshold. In this case, the wall-clock time may be easily estimated from the total number of requests until failure and the average request rate.

After introducing preliminary definitions, next step is to obtain the pilot sample as discussed in Section 3. Table 2 summarizes the traditional plan adopted for this case study. The number of replications is related to the pilot sample.

As experimentally verified in Ref. 15, the request rate has no influence on the httpd aging, hence, we use the highest rate supported by the SUT in each stress level. We stop a test only when the size of the httpd crosses

Table 2. Experimental plan.

Stress loading		Allocation	
Level	Page size (kB)	Proportion π	Replications n_{ALT}
Use	200		
S1	400	1/3	7
S2	600	1/3	7
S3	800	1/3	7

the 100-megabyte threshold. Table 3 shows the pilot sample of accelerated failure times (TTF).

From the samples of failure times (TTF), we tested the relevant probability distributions (see Section 3). The criteria used to build the best-fit ranking were the log-likelihood function (Lk), and the Pearson's linear correlation coefficient (ρ), whose parameter estimation methods were MLE, and LSE, respectively. The goodness-of-fit (GOF) test results for the considered models are shown in Table 4.

The Weibull probability distribution showed the best fit for the three accelerated lifetime data sets as demonstrated by the numerical assessment (Lk, and ρ). According to the algorithm described in Section 3, the data set must be transformed using the natural logarithm ($\log_e$). Table 5 shows the natural logarithms of pilot sample.

From the transformed dataset, we calculate the number of replications solving Eqs. (7) to (11). As a result, we obtained $n_{ALT} = 11$, which means

Table 3.   Sample of time to failure.

TTF ($S1$)	TTF ($S2$)	TTF ($S3$)
84	34	20
86	36	21
88	37	22
93	38	23
95	38	23
95	39	23
97	40	24

Table 4.   Model fitting for accelerated failure times.

Stress level	Model	GOF		Best-fit ranking
		Lk	ρ (%)	
$S1$	Weibull	−20.5086	96.75	1st
	Lognormal	−20.8627	95.98	2nd
	Exponential	−38.5870	−74.38	3rd
$S2$	Weibull	−13.9211	99.31	1st
	Lognormal	−14.3368	97.60	2nd
	Exponential	−32.3570	−74.53	3rd
$S3$	Weibull	−11.2420	97.65	1st
	Lognormal	−11.8037	95.33	2nd
	Exponential	−28.7276	−73.96	3rd

Table 5.   Transformed pilot sample.

TTF ($S1$)	TTF ($S2$)	TTF ($S3$)
1.9243	1.5315	1.3010
1.9345	1.5563	1.3222
1.9445	1.5682	1.3424
1.9685	1.5798	1.3617
1.9777	1.5798	1.3617
1.9777	1.5911	1.3617
1.9868	1.6021	1.3802

eleven replications for the ALT experiment. Because the size of pilot sample is bigger than n_{ALT}, we decide to use the pilot sample to estimate the metrics of interest. The Weibull distribution also showed the best fit for the transformed sample of failure times, therefore we use it in conjunction with the IPL model to create our life-stress relationship model.

According to Ref. 17, the Weibull distribution may adopt the same parameterization structure shown in (3), where $\sigma = 1/\beta$ is the scale parameter, and $\mu = \log(\eta)$ is the location parameter. Hence, the assumption of same scale parameter across the stress levels must be evaluated on the estimated values of β after fitting the Weibull model to the three samples of failure times. We verified that the three beta values are inside the confidence interval calculated for each sample, thus satisfying the assumption of scale invariance. Table 6 presents the estimates for Weibull parameters, obtained through the maximum likelihood (ML) parameter estimation method.

In order to have the IPL-Weibull relationship model, we make $\eta = L(s)$ (see Eq. (1)). As a result we obtain (11), which is the IPL-Weibull *pdf*

$$f(t, s) = \beta k s^w (k s^w t)^{\beta - 1} e^{-(k s^w t)^\beta}. \qquad (12)$$

Table 6.   Estimated Weibull parameters.

Stress	Parameter	ML Estimate	CI (90%) Lower	Upper
$S1$	β_1	24.0889	14.3941	40.3134
	η_1	93.3175	90.8149	95.8891
$S2$	β_2	25.0000	15.2671	40.9377
	η_2	38.2697	37.2791	39.2865
$S3$	β_3	22.0825	13.3316	36.5775
	η_3	22.8595	22.1925	23.5465

Thus, (12) can be directly derived from (11) and used to estimate the mean time to failure, MTTF, of the SUT for any use rate:

$$MTTF = \frac{1}{ks^w} \cdot \Gamma\left(\frac{1}{\beta} + 1\right),\tag{13}$$

where Γ is the Gamma function.[26]

Table 7 shows the estimates for the IPL-Weibull model parameters.

Using the estimated parameters for the IPL-Weibull model, we calculate the mean life of the SUT for its use condition. Figure 6 presents the stress-life relationship projected for all stress levels with the abscissa values beginning at the use condition level. The ordinate values correspond to the time-to-failure in batches of 100 requests targeting pages dynamically created (aging factor). The intersection point between the y-axis, and the

Table 7. Estimated IPL-Weibull parameters.

| Parameter | ML Estimate | CI (90%) | |
		Lower	Upper
k	5.7869E-8	3.7257E-8	8.9885E-8
w	2.0340	1.9646	2.1034
β	18.9434	13.5270	26.5286

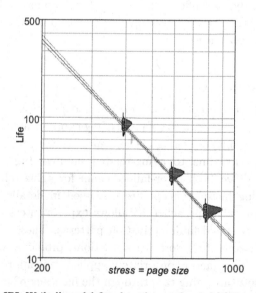

Fig. 6. IPL-Weibull model fitted to the accelerated lifetime dataset.

line obtained by linearizing the IPL-Weibull model is the MTTF estimated for the use rate condition level.

In order to evaluate the accuracy of the estimates, we run one experiment with seven replications using the page size equals to 200 kB. This setup refers to the SUT operating in its normal regime. High accuracy of our fitted IPL-Weibull model is confirmed with the observed MTTF (343.57) falling within the estimated 90% confidence intervals (337.92–395.28).

As stated in Section 1, the main goal of using ALT in software aging experiments is to reduce the observation time required to collect aging-related failure times. We calculate the reduction factor obtained with the aging acceleration method for the total experimentation time. First, we compute the total time spent to execute all tests (replications) for all stress levels. As explained before, we collected the aging-related failure times in terms of bunches of 100 requests. Thus, based on the sample presented in Table 3, the experimentation time, in number of requests, is 105,600 requests. The mean time to aging-related failures observed for the experiment in use condition (non-accelerated aging) was 343.57, then resulting in 34,357 requests. Considering the same number of replications (twenty one) used in the case study, we have a total experimentation time of 7,214.97 bunches of 100 requests, or 721,497 requests. Therefore, in this experimental study, ALT offers a reduction of approximately 7 (seven) times the time required to obtain a sample of twenty-one failure times in use condition (w/o acceleration).

5. Final Remarks

In this chapter we present the theoretical and practical aspects of using accelerated life tests applied to software aging experiments. We discuss how the occurrence of software aging effects can be accelerated, in a controlled way, to reduce the time to aging-related failures. Important concepts such as the modified fault-error-failure chain for aging-related events, as well as the aging factor concept are discussed in details. We present a design-of-experiment approach that allows experimenters to identify the significant aging-related fault activation patterns. These patterns can be individual operations of the software operational profile, as well as system's operational modes. When appropriately applied, this approach allows fine control of the software aging rate through the handling of the aging factors into the workload, or even through environmental variables. The case study section provide a complete example of how to apply the ALT method in studies related to the reliability analysis of systems suffering from software aging. This approach offers a reduction in the time to observe aging-related failures, as verified in the case study where we obtained a reduction of seven times the time required to get a sample of twenty-one failure times.

Several recent studies in experimental software aging and rejuvenation research are being conducted to investigate the aging effects not only in different application architectures but also in operating system kernel level. Specially, this second line of investigation is very important, since aging effects inside the OS kernel affect significantly the entire system, differently than in application level that usually is confined to the affected application address space. Initial studies in OS aging are mainly investigating the aging effects in the virtual memory (e.g., fragmentation), device driver subsystems (e.g., memory leaks), and file system metadata (e.g., fragmentation). In these cases, advanced kernel instrumentations are required to monitor and log kernel events in real time. Specially in embedded systems,[14] the impact of OS aging effects are very significant, since the system's resource such as main memory, storage space, etc., are very limited.

References

1. Avižienis, A., Laprie, J.-C., Randell, B., Landwehr, C.: Basic concepts and taxonomy of dependable and secure computing, *IEEE Transactions on Dependable and Secure Computing*, 1(1), 2004, pp. 11–33.
2. Avritzer, A., Weyuker, E.J.: Monitoring smoothly degrading systems for increased dependability, *Empirical Software Engineering*, 2(1), 1997, pp. 59–77.
3. Bao, Y., Sun, X., Trivedi, K.S.: A workload-based analysis of software aging and rejuvenation, *IEEE Transactions on Reliability*, 55(3), 2005, pp. 541–548.
4. Chillarege, R., Goswani, K., Devarakonda, M.: Experiment illustrating failure acceleration and error propagation in fault-injection, in *Proc. International Symposium on Software Reliability Engineering*, 2002.
5. Cisco Systems, Inc., Cisco security advisory: Cisco Catalyst memory leak vulnerability, Document ID: 13618, 2001. URL = http://www.cisco.com/warp/public/707/cisco-sa-20001206-catalyst-memleak.shtml.
6. Dohi, T., Goševa-Popstojanova, K., Trivedi, K.S.: Estimating software rejuvenation schedules in high assurance systems, *Computer Journal*, 44(6), 2001, pp. 473–482.
7. Ehrlich, W., Nair, V.N., Alam, M.S., Chen, W.H., Engel, M.: Software reliability assessment using accelerated testing methods, *Journal of the Royal Statistical Society*, 47(1), 1998, pp. 15–30.

8. Grottke, M., Li, L., Vaidyanathan, K., Trivedi, K.S.: Analysis of software aging in a web server, *IEEE Transactions on Reliability*, 55(3), 2006, pp. 411–420.

9. Grottke, M., Trivedi, K.S.: Software faults, software aging, and software rejuvenation, *Journal of the Reliability Association of Japan*, 27(7), 2005, pp. 425–438.

10. Grottke, M., Trivedi, K.S.: Fighting bugs: Remove, retry, replicate and rejuvenate, *IEEE Computer*, 40(2), 2007, pp. 107–109.

11. Grottke, M., Matias, R., Trivedi, K.: The fundamentals of software aging, In Proc of Workshop on Software Aging and Rejuvenation, in conjunction with *IEEE International Symposium on Software Reliability Engineering*. 2008.

12. Huang, Y., Kintala, C., Kolettis, N., Fulton, N.: Software rejuvenation: analysis, module and applications, In *Proc. Twenty-Fifth International Symposium on Fault-Tolerant Computing*, pp. 381–390, 1995.

13. Jain, R.: The art of computer systems performance analysis: techniques for experimental design, measurement, simulation, and modeling. John Wiley and Sons, 1991.

14. Kintala, C.: Software rejuvenation in embedded systems, *Journal of Automata, Languages and Combinatorics*, 14(1), 2009, pp. 63–73.

15. Matias, R., Freitas, P.J.: An experimental study on software aging and rejuvenation in web servers, In *Proc. 30th Annual International Computer Software and Applications Conference*, Vol. 1, 2006, pp. 189–196.

16. Mettas, A.: Understanding accelerated life testing analysis, In *Proc. of International Reliability Symposium*, 2003, 1–16.

17. Meeker, W.Q., Escobar, L.A.: Statistical methods for reliability data. New York: Wiley, 1998.

18. Montgomery, D.C.: Design and analysis of experiments, 6th ed. John Wiley and Sons, 2005.

19. Musa, J.D.: Operational profiles in software reliability engineering, *IEEE Software*, 10(2), 1993, pp. 14–32.

20. Musa, J.D.: Software Reliability Engineering, McGraw-Hill, 1999.

21. Nelson, B.N.: Accelerated testing: statistical method, test plans, and data analysis, New Jersey: Wiley, 2004.

22. Oliveira, V.R.B., Colosimo, E.A.: Comparison of methods to estimate the time-to-failure distribution in degradation tests, *Quality and Reliability Engineering International*, 20(4), 2004, pp. 363–373.

23. Shereshevsky, M., Crowell, J., Cukic, B., Gandikota, V., Liu, Y.: Software aging and multifractality of memory re-sources, In *Proc. IEEE International Conference on Dependable Systems and Networks*, 2003, pp. 721–730.

24. Silva, L., Madeira, H., Silva, J.G.: Software aging and rejuvenation in a SOAP-based server, In *Proc. Fifth IEEE International Symposium on Network Computing and Ap-plications*, 2006, pp. 56–65.

25. Tobias, P., Trindade, D.: Applied Reliability, 2nd ed. Kluwer Academic Publishers, Boston, 1995.

26. Trivedi, K.S.: Probability and Statistics with Reliability, Queuing, and Computer Science Applications, John Wiley and Sons, New York, 2001.

27. Vaidyanathan, K., Trivedi, K.S.: A comprehensive model for software rejuvenation, *IEEE Transactions on Dependable and Secure Computing*, 2(2), 2005, pp. 124–137.

28. "http server project", Apache Software Foundation [Online]. Available at: http://httpd.apache.org.